Multifunctionality in Agriculture

WHAT ROLE FOR PRIVATE INITIATIVES?

OECD

ORGANISATION FOR ECONOMIC CO-OPERATION AND DEVELOPMENT

ORGANISATION FOR ECONOMIC CO-OPERATION AND DEVELOPMENT

The OECD is a unique forum where the governments of 30 democracies work together to address the economic, social and environmental challenges of globalisation. The OECD is also at the forefront of efforts to understand and to help governments respond to new developments and concerns, such as corporate governance, the information economy and the challenges of an ageing population. The Organisation provides a setting where governments can compare policy experiences, seek answers to common problems, identify good practice and work to co-ordinate domestic and international policies.

The OECD member countries are: Australia, Austria, Belgium, Canada, the Czech Republic, Denmark, Finland, France, Germany, Greece, Hungary, Iceland, Ireland, Italy, Japan, Korea, Luxembourg, Mexico, the Netherlands, New Zealand, Norway, Poland, Portugal, the Slovak Republic, Spain, Sweden, Switzerland, Turkey, the United Kingdom and the United States. The Commission of the European Communities takes part in the work of the OECD.

OECD Publishing disseminates widely the results of the Organisation's statistics gathering and research on economic, social and environmental issues, as well as the conventions, guidelines and standards agreed by its members.

Publié en français sous le titre :
La multifonctionnalité dans l'agriculture
QUEL RÔLE POUR LE SECTEUR PRIVÉ ?

Foreword

OECD has developed and published an analytical framework and guidelines intended to help governments find the best solutions to problems related to agricultural externalities and public goods *Multifunctionality: Towards an Analytical Framework* (2001) and *Multifunctionality: The Policy Implications* (2003). One avenue to be explored in the search for the most efficient solutions relates to non-governmental approaches. This is the subject of the present report. The use of market mechanisms, private transactions, and voluntary approaches are examined. Information on non-governmental approaches that have been implemented to date, is provided, highlighting the institutional settings and the policy implications of these approaches when compared to governmental approaches.

This study was carried out by the Policies, Trade and Adjustment Division of Directorate for Food, Agriculture and Fisheries, with Jun Shobu as the principal author. Valuable contributions were provided by Darryl Jones also from the Directorate for Food, Agriculture and Fisheries (Case Study 13), and consultants Beatriz E. Velázquez of the Istituto Nazionale di Economia Agraria (INEA) (Case Study 1), Gerald J. Pruckner of the Department of Economics and Statistics, University of Innsbruck in Austria (Case Study 5), Ståle Navrud of the Department of Economics and Resource Management, Agricultural University of Norway (Case Study 6), Simon Swaffield of Lincoln University, New Zealand (Case Study 9), Gilles Grolleau of UMR INRA-ENESAD (Case Study 10), and the New Zealand government (Case Study 12). Other colleagues from the Directorate and Ken Whitley, Wildcare Dairy Group Ltd., also provided useful comments. The report was declassified by the Working Party on Agricultural Policies and Markets of the Committee for Agriculture in September 2005.

Table of contents

Foreword... 3

Part I. Synthesis Report 7

    Assessment of Non-Governmental Approaches: Non-Commodity Outputs
(NCOs)... 13

    Assessment of Non-Governmental Approaches: Negative Effects (NEs) 23

    Conclusions... 28

    References... 30

Part II. Case Studies... 33

    1.   Agritourism in Italy (Type P1).. 35

    2.   Market price premiums for milk in the United Kingdom (Type P2) 41

    3.   Conservation Trust: The National Trust in the United Kingdom (Type P3)........... 47

    4.   Easement arrangements in the United States (Type P3) 52

    5.   Agritourism and landscape conservation program in Austria (Type P3)............. 57

    6.   Tourist train in Norway (Type P3)... 63

    7.   Consumer movement: *Chisan-chishou* in Japan (Type P4) 69

    8.   Voluntary flood mitigation (Type P5)... 75

    9.   Voluntary conservation of biodiversity and landscapes on Banks Peninsula,
New Zealand (Type P5) 79

    10.   Direct transaction: a case of mineral water in France (Type N1)........................ 83

    11.   Voluntary approaches (Type N2) 89

    12.   Sustainable Winegrowing New Zealand® (Type N2) 95

    13.   Tradeable manure/animal quotas in the Netherlands (Type N3)......................... 99

    14.   Wetland Mitigation Banking (Type N1) 113

    References... 121

Part I.

Synthesis Report

Abstract

Part I explores non-governmental approaches to dealing with externality problems associated with agricultural production, with the aid of concepts and ideas drawn from institutional economics. A typology is developed that covers both positive externalities (non-commodity outputs) and negative effects of agriculture. Different non-governmental approaches are examined from the perspectives of economic efficiency, effectiveness, equity, and stability in comparison with government approaches, based on a large number of case studies which are reproduced in Part II of the report.

OECD governments are committed to a process of agricultural policy reform that aims to improve efficiency and minimize production and trade distortions arising from existing instruments and to develop new approaches. Against this background, this study seeks to identify situations in which non-governmental approaches to dealing with positive and negative externalities of agriculture might be more efficient, more stable or more equitable than direct government intervention. The exploration of non-governmental approaches (NGAs) arose from work on the multifunctionality of agriculture [*Multifunctionality: Towards an Analytical Framework*, 2001 and *Multifunctionality: The Policy Implications, 2003*]. An explicit recommendation from that work was that, in the presence of market failure, non-governmental options such as market creation or voluntary provision should be explored before any decision to proceed to direct government intervention.

Objectives

Problems related to positive and negative externalities and to public goods cannot be solved by market forces alone or exclusively through government interventions that can be costly and inefficient (OECD, 2001a). The OECD has developed policy guidelines for governments for multifunctionality and this project goes a step further in seeking ways to encourage NGAs for the provision of NCOs or the reduction of NEs through the use of market mechanisms, and the promotion of private transactions and voluntary approaches. The solutions envisaged may involve a minimum level of government intervention and some associated budget costs. A general framework has been developed and is based on comparative case studies by which policy makers can analyse interactions among institutions, and thus establish policies to encourage NGAs where this is the most efficient solution.

Approaches

Generally, both NCOs and NEs, due to their public good or externality characteristics, are not traded in markets and their provision/reduction is not remunerated in the absence of special arrangements. Beneficiaries of the goods cannot be segregated or segregation is not feasible due to high costs (non-excludability), with the result that there is no valuation of the good in question. Financial incentive arrangements that enable the trading of NCO/NE include transactions in agricultural products and NCOs combined, and specifying property rights concerning either the right to pollute or to mitigate pollution. Such arrangements work to overcome or mitigate the problems of non-excludability or the absence of valuation, and in some cases act as a mechanism to facilitate NCO/NE transactions.

This project is comprised of three steps. The first step was to establish an analytical framework based on a literature review. The second step consisted of providing an in-depth analysis through selected case studies as presented in the annex of this report. The final step synthesises these works.

The analytical framework provided a typology of NGAs for both NCOs and NEs (Table I.1) and an analytical framework that included institutional economics and transaction cost analysis. But due to methodological difficulties and data unavailability, no attempt was made to undertake quantitative transaction cost analysis. NGAs for NCOs were classified into market, club, and voluntary provision and then further divided according to whether supply was separate or joint with agricultural commodities. This generated five types of NCOs. NGAs for NEs were divided into three types: market-based approaches, direct transactions, and farmer-led voluntary approaches.

Table I.1. Typology of NGAs

Type N°	Name of Types	Features	Typical Case	Implemented Case Studies
Type P1	Market provision (Consumers: Individuals)	Discrete supply	Agri-tourism	• Agritourism in Italy
Type P2		Joint supply with commodities	Market price premiums	• A brand of milk production and sale in the UK
Type P3	Club provision (Consumers: Clubs or organizations)	Discrete supply	Conservation trusts	• The National Trust in the UK • Easement arrangement in the US • Agritourism in Austria • Tourist train in Norway
Type P4		Joint supply with commodities	Community Supported Agriculture	• Consumer movement in Japan
Type P5	Voluntary provision	No transactions		• Voluntary flood control in Japan • Biodiversity and landscape Conservation in New Zealand
Type N1	Direct transactions	Negotiations and transactions between farmers and sufferers		• A mineral water bottler in France
Type N2	Farmer-led voluntary approaches	No transactions		• Findings and lessons are drawn from earlier OECD work • Wine growing program in New Zealand
Type N3	Market-based approaches	Trading permits/credits in markets		• Manure trading in the Netherlands • Mitigation banking in the US

Non-governmental approaches (NGAs)

Two broad schools, *i.e.* Pigovian and Coasian, have dominated economic analysis of externality problems and resultant policy designs (Garvie, 1999). The Pigovian school seeks to centralize solutions through taxes and subsidies, but regulators have imperfect knowledge about the costs and benefits of pollution abatement which results in errors in the targets and rule-setting process and, consequently, welfare losses (*Ibid*). In contrast, the Coasian school prescribes that government intervention should be restricted to the definition of property rights, and that under well-defined property rights, pure non-governmental solutions that are in theory Pareto optimal will be reached through decentralized negotiations. The NGAs studied here relates to the search for the latter type of solution where government plays an indirect role in the negotiations.

For NCOs, NGAs are defined either as those that involve contracting between private entities for the supply of NCOs with payment from consumers (demanders) to farmers (suppliers), or as those involving suppliers voluntarily providing NCOs without compensation. For the former type, consumers in this project may be individuals or non-governmental organizations (NGOs) that represent a group of consumers. Public entities may take part in the contracting in various forms including the establishment of transaction rules or provision of technical or even financial assistance. However, these particular public entities never directly contract nor oversee the contracting. Voluntary provisions may involve contracts with consumers, but no pecuniary transactions are involved. Further, voluntary provision is defined as involving additional resource use. There is no policy issue where NCO provision occurs without any additional resources as part of the normal activity on a farm. In this regard, for the former type, *i.e.* market or club provision, whether additional inputs are used or not is not an issue, although, in practice, production of NCOs is likely to require additional resources.

Negative effects are considered as "bads". Society wishes to reduce them, but the problem is who bears the costs and how. A conventional approach is a control-and-command (CAC) one which directly regulates pollution, imposing a uniform burden on all polluters by government authorities. Compliance cost may be high and innovation for the reduction of pollution may be stifled. This study seeks other solutions under the Coasian approach where pollution reduction measures are determined by the polluter and compliance cost is low relative to the CAC approach.

NGAs for reducing NEs are direct transactions between polluters and pollutees. However, such direct transactions rarely take place in view of the incomplete property rights and/or agriculture's non-point source pollution.[1] Therefore, NGAs in this project are defined more widely, extending to market-based and to voluntary approaches where private initiatives and market forces play a central role and are alternative solutions to government regulations. These are incentive-based approaches that provide a level of flexibility with the purpose of rewarding environment-enhancing behaviour absent in CAC approaches, resulting in the reduction of compliance costs, efficient resource allocation and innovation in environmental technologies (Randall and Taylor, 2000). They include market-based approaches such as tradeable pollution permits/credits/quotas, in which market forces determine the allocation of permits/credits/quotas to polluters or third parties who can achieve policy targets in a more efficient manner than in a CAC regime; governments develop and enforce the targets and market rules. In this sense, this scheme is an amalgam of private initiatives and government regulatory intervention. Voluntary approaches to cope with environmental problems as an alternative to mandatory approaches are on the rise in various sectors (Wu and Babcock, 1999).[2] Voluntary approaches in the agricultural sector involve farmers or industry-led activities to reduce NEs, motivated by long-term financial benefits or fear of solutions imposed by a regulatory authority. Government financial or technical supports can be found to catalyze them, but are not substantial. Many examples of voluntary approaches can be seen as a form of co-operative activity. This study reviews and examines these approaches from the viewpoints of cost-effectiveness and efficiency, and seeks ways to encourage private initiatives, where appropriate.

Case studies

A set of comparative case studies provide practical examples of solutions to the issues of non-remuneration of NCO provision and the difficulty in reducing NEs. Institutional and organizational settings are examined. Each case study recognizes the differences in settings prevailing in each country and tries to draw lessons relevant to other societies, but does not offer general policy recommendations nor an evaluation of the settings. Jointness between agricultural production and NCOs and NEs is examined in all the case studies, but the degree of jointness is outside the scope of this project.

The information on actual cases was collected from the literature, from relevant websites, and some was provided by delegations. The list of cases examined is presented in Table I.2 and the relevant reports are found in the Annex. One to three cases are selected from each of the eight types, categorized according to the analytical framework. It should be noted that agritourism (Type P1) in particular can be observed in many or all OECD countries, but that other types are limited to a few or only one country. Thus, this selection does not reflect the actual distributional pattern of non-governmental efforts in OECD countries. The selection criterion was to examine as many types as possible, because they have salient institutional settings and offer broader policy choices. Among the cases studied, some have long operational experience and thus have had certain impacts; others have just started and cannot be evaluated fully. In addition to elements that are keys to success, the difficulties encountered are also mentioned as these can also be instructive.

Table I.2. List of Case Studies and Brief Descriptions

Type Number	Serial Number	Implemented Case Studies	Brief Descriptions
Type P1	CS 1	• Agritourism in Italy	The legal setting in Italy promotes agritourism. Cultural heritage and landscape are addressed in this case, but there is no evidence that the NCO values are priced and transacted.
Type P2	CS 2	• A brand of milk production and sale in the UK	A price premium is added to the purchase price of milk, with the amount of that premium being used for conservation work by individual farmers supplying the milk.
Type P3	CS 3	• The National Trust in the UK	The National Trust collects funds from its supporters and invests them to conserve the resources in countryside, including its-owned farmlands where the tenants implement conservation work agreed with the trust.
	CS 4	• Easement arrangement in the US	Agricultural conservation easement is an agreement between a landowner (farmer) and an NGO concerning permanent release of development rights. The landowner receives a lump sum payment or tax deduction as compensation.
	CS 5	• Agritourism in Austria	A private community program taxes tourists who stay the community and pays farmers who observe landscape cultivation guidelines.
	CS 6	• Tourist train in Norway	A group of farmers operating agriculture and conserving farm landscape along railway tracks are paid by private companies and a local community.
Type P4	CS 7	• Consumer movement in Japan	A consumer movement aims to promote the consumption of locally-produced foods and to conserve local agriculture. No clear NCO transactions seem to be involved.
Type P5	CS 8	• Voluntary flood control in Japan	Paddy rice growers voluntarily mitigate downstream floods by delaying the rainwater discharge in cases of heavy rain, motivated by a sense of community solidarity. There is no pecuniary remuneration for the farmers' action.
	CS 9	• Biodiversity and landscape Conservation in New Zealand	A community-based trust works with landowners (including farmers) to voluntarily protect biodiversity and landscape by means of conservation covenants.
Type N1	CS 10	• A mineral water bottler in France	As a result of private negotiations between polluters (farmers) and a pollutee (a private firm), the latter agreed to pay the former who perform extensive agriculture at the upstream of a mineral water spring, by which nutrient discharge polluting underground water can be mitigated.
Type N2	CS 11	• Findings and lessons are drawn from earlier OECD work	Four sub-case studies that applied farmers-led voluntary approaches are examined. The common focus is soil degradation leading to future financial loss, which is a key motivation for participating farmers.
	CS 12	• Wine growing program in New Zealand	An industry-led sustainable wine growing program urges farmer's voluntary participation. The program not only provides cost-saving technologies mainly by reducing the costs of pesticides to reduce nutrient pollution and soil degradation but also allow the use of logo certifying sustainable cultivation.
Type N3	CS 13	• Manure trading in the Netherlands	Tradeable quota for manure production and animals were implemented to place a cap on manure nutrients, with trading rules to achieve a better environmental distribution of livestock.
	CS 14	• Mitigation banking in the US	Wetland developers purchase mitigation credits from a mitigation bank that create the credits through the restoration, construction or enhancement of wetlands. By this, no net loss of wetlands is pursued.

Assessment of Non-Governmental Approaches: Non-Commodity Outputs (NCOs)

Before proceeding with an assessment of NGAs, a few relevant economic aspects of NCOs relating to general public good properties and those specific to NCOs, are revisited. Findings from the case studies are then itemized, followed by a more in-depth examination to facilitate understanding of NGAs in so far as NCOs are concerned, how transaction mechanisms are formed, and the issues involved in NCO transactions. Lastly, NGAs are assessed in a more general manner and conclusions are presented, including on the role of governments with respect to optimal NCO provision.

Some economic aspects of NCOs

Public good characteristics, *i.e.* non-excludability and non-rivalry, prevent NCOs from being transacted in markets, although NCOs do not all necessarily have both of these characteristics. Non-excludability is a key condition explaining why transactions do not occur. Running an excludability mechanism is prohibitively costly or unfeasible for public goods in general. Where institutional arrangements or technological innovations can reduce excludability costs, it becomes easier for goods to be subject to transaction. Non-rivalry can be characterised as the marginal cost of providing a good to an additional individual being very low or almost nil (Stiglitz, 1998). Unlike non-excludability, it is not a factor that prevents transactions from taking place. For example, education and water supply services that are excludable but in general non-rival can be transacted in a market. However, non-rivalry relates to efficiency, as efficient resource allocation occurs when goods are traded at marginal cost and equilibrium between supply and demand is created. There is a prima facie case for governments to supply goods for which the marginal cost is minimal or transaction costs are significant. Otherwise, there might be significant welfare losses (*ibid*)[3].

In addition to the generic public good characteristics mentioned above, NCOs have their own specific features that make their examination more complex. The production of commodities may result in the simultaneous production of several NCOs, *i.e.* environmental services (such as landscape, biodiversity, and wildlife habitats), food security, and/or land conservation. Each NCO has its unique public good features. Some are closer to pure public goods and others have club good or impure public good characteristics, depending on their degree of non-excludability and non-rivalry. Moreover, the number of beneficiaries is different, and affects the nature of public goods and the likelihood of private provision. Therefore, it can be assumed that measures to generate private transactions for an NCO will differ, giving rise to the question as to how actual NGAs deal with this problem.

Findings from case studies

A number of findings are drawn from each case study. The following are common to most of the cases and are therefore relevant to a general assessment of NGAs.

- Most NCOs traded under NGAs can be categorized as environmental services, such as landscape and wildlife habitats. It seems to be difficult for other NCOs to be transacted in private initiatives.

- Each case delimits the farmers or farmlands to participate in an NCO transaction project so that the NCO to be transacted is also delimited. This is the first step for NCO transactions to occur.

- No case has a clear mechanism that intentionally excludes free-riders.

- Two potentially conflicting objectives, *i.e.* the promotion of environment conservation and agricultural production[4], are reconciled. This seems to be an important basis for successful NCO transactions.

- A range of different valuation and pricing measures serve NCO transactions.

- There are various types of transactions in which measures to lower transaction costs are taken.

- A range of government interventions are found in most of the cases.

In the sections that follow each of these findings is examined in-depth and the resulting analysis is the basis for the final assessment.

NCOs traded and not traded

Although there are multiple NCOs produced from agricultural practices and its outputs, the majority of NCOs traded under NGAs are agri-environmental in nature. The reasons for this seem to be the following:

- Consumer's willingness to pay is relatively higher than for other NCOs. This provides financial resources that enable private agents to undertake transactions.

- Agri-environmental services are commonly and broadly demanded in most, if not all, societies.

- There are a number of working environment NGOs in place that act as catalysts for NCO transactions.

Environment services traded under NGAs can be divided into two groups: those that have high biological or landscape values, and those with more ordinary environment characteristics. The former is found in the case of agritourism in Italy (CS1: the number of this case study shown in Table 2), the National Trust (CS3), the Norwegian tourist train (CS6), and the landscape conservation program in Austria (CS5). Their scarcity generates a high (tourism) value, and thus attracts many consumers. The latter group is found in the cases of the price premium for milk in the UK (CS2), easement arrangements in the US (CS4), and the consumer movement in Japan (CS7). These cases do not involve scarcity nor a high value, so special mechanisms to attract consumers to pay for the ordinary environment have to be developed. The case of the market price premium (CS2) minimizes the payment as much as possible, 5 pence per bottle of milk, to encourage as many consumers as possible to participate in the scheme. For the Japanese and US cases, their targeted objectives are unique. They target rural consumers and the conservation of the nearby environment in the Japanese case (CS7), or protect agricultural lands (many of which are near urban areas) from development in the US case (CS4).

On the other hand, other types of NCOs do not seem to lend themselves to NGAs as follows:

- *Food security*: While the case of community supported agriculture (CSA) might seem to be motivated by the desire of the participating consumers to improve their sense of food security through exclusive food provision, in fact the belief that they are receiving safer food and a more secure food supply while also preserving or revitalising local agriculture would seem to be more strongly motivating factors. More generally, the absence of non-governmental approaches in the domain of food security may be explained by the fact that food security is not an issue or is not perceived to be an issue. Where food security is perceived to be an issue, the paucity of NGAs is most likely explained by the pure public good nature of food security, which means that it will be most efficiently provided by governments. Whether domestic production, stockholding, imports or some combination of these is the most efficient way to provide food security in any given situation is a separate question beyond the scope of this report.

- *Land conservation*: This consists of flood control, soil conservation, landslide prevention and ground water recharge. People have long benefited from these services without payment and, further, these services have been provided by agricultural practices without additional cost. However, it is also clear that if free provision in association with agricultural production was threatened, that the impure public good characteristics of these NCOs are such that beneficiaries could be identified and organised to pay for the services. In other words, it would be possible to organise exclusion and paying mechanisms in these cases. A case of flood control was studied in Japan as a voluntary provision, where no pecuniary transactions occur.

Delimitation of NCOs to be transacted

NCO services are supplied by a discrete or a large number of contiguous farming lots, or all farmlands in a country or region. To deliver an NCO service to a market, it is necessary for this NCO service to be institutionally confined and isolated from others (Barthelemy and Nieddu, 2004) by delimiting farmers or farmlands qualified to participate in a project. It is then ready to be valued by potential demanders and actual transactions can proceed. This confinement determines the size of the project and market.

In the case involving the UK market price premium scheme (CS2), suppliers delimit the NCO by limiting the number of participating farmers. Thus, an NCO, the conservation of natural habitat in this case, is restricted to that produced from their farmlands. In the case of the National Trust (CS3), targeted NCOs, *i.e.* farmlands to be conserved, are determined by council members which represent all the members. Unlike the UK case, geographical and administrative boundaries are applied to the Norwegian (CS6) and Austrian cases (CS5), respectively. In Norway, a targeted NCO is a landscape in a valley. For the Austrian case, farmers and the tourist industry in a local town participate in a landscape conservation program. The targeted NCO is thus limited to the one produced in the locality.

Exclusion mechanisms

In theory, if demanders can receive the benefits of a service or good regardless of whether they pay for it, there will be little motivation to reveal their reservation price, which is essential information to determine the value of the service or good. Without a proper exclusion mechanism, the demanders can behave as free-riders and accordingly no transactions occur.

There is no clear exclusion mechanism for most of the cases examined, and consumers pay for NCO services even though they seem to be aware of the existence of free-riders. The reasons for this attitude may be as follows:

- NCO consumers seem to feel that payment for environmental services is charitable because the services benefit an entire region or country.

- Many environment services have non-use value, which means that consumers rarely visit the sites, but wish to conserve them for the general public or the next generation.

- For some cases, the payment by the consumer is small relative to the private benefit, and thus too small to justify excluding free-riders.

- The cost of setting up and maintaining exclusion mechanisms is very high.

As can be seen in the Norwegian (CS6) and Austrian (CS5) cases targeted NCOs can be geographically enclosed or confined to a community (or administrative unit). There are, of course, free-riders but they have a disadvantage, for example, with regard to their location from where they can see landscape. Meanwhile payers can enjoy it from the best location. In these cases, soft exclusion mechanisms work. The Norwegian case exemplifies a physical mechanism while the other an institutional one.

Objectives for NCO conservation

The pursuit of higher quality environmental services may conflict with that of profitable agriculture, especially where intensive production is concerned. These two conflicting objectives, *i.e.* profitable production and environmental conservation, are institutionally reconciled and pursued simultaneously in the cases examined. The farmers' motivation to participate in NCO transaction schemes is maintained through remuneration or compensation.

In this respect, the case of the market price premium scheme in the UK (CS2) is a good example in that a balance between commodity production and environment services is found by converting 10% of all participating farmlands to environmental use. This rule not only resolves the problem of environment *vs.* production, but also offers a win-win solution. The more premium milk sold, the more such milk production is promoted, and the more the environment adjacent to the production areas is conserved. Other cases pursue compromise through individual negotiations between two parties who want to protect the environment and promote production. This solution creates transaction costs but is suitable for heterogeneous farming conditions.

Valuation and pricing of NCOs

NCO prices are not determined through a bargaining process in a competitive market, but basically through negotiations or offers from suppliers without negotiation. For a scheme to work, the price level is important. In the absence of a competitive market, price determination should reflect costs of maintaining and operationalising schemes as well as the potential demand for a specific NCO.

Consumers in the cases of the National Trust (CS3), the market price premiums (CS2), the tourist train (CS6), and the land conservation program (CS5) want higher environment quality than that produced by ordinary agriculture. The price of the NCOs is thus basically determined to meet the additional farmers' costs or lost yields, incurred by the creation and maintenance of hedges, restoration of woodlands or wetlands, clearance of bushes, or limiting livestock density. If consumers demand a higher level of NCO quality, more investment is required. An alternative pricing scheme involves consumers paying farmers for a commitment to continue farming (CS4). The amount is based on the difference in the value of the property before and after placing the easement on particular parcels of land. The payment is made as a lump sum or a tax reduction over the course of several years. This pricing scheme is unique in that the price is determined objectively. But, as the price is specific to each plot the pricing exercise creates large transaction costs. The consumer movement in Japan (CS7) for the maintenance of local agriculture does not involve any payment. Thus, no valuation is needed.

Transaction mechanisms and transaction costs

Coase (1960) stated that "[i]n order to carry out a market transaction it is necessary to discover who it is that one wishes to deal with, to inform people that one wishes to deal with and on what terms, to conduct negotiations leading up to a bargain, to draw up the contract, to undertake the inspection needed to make sure that the terms of the contract are being observed, and so on." These elements are necessary more or less in any transaction process and incur costs. With respect to actual NCO transactions, some are less important, or least costly.

The crucial point for successful transactions to occur is that aggregate transaction costs are sufficiently low, although there is no predetermined threshold. Transaction costs can be broken down into: information, negotiation, and enforcement costs (Furubotn and Richter, 1998). For the actual NCO transactions observed in this study, there is no uniform tendency for specific categories of costs to be high or low among all institutional forms. Their costs depend on the individual institutions and the transaction mechanisms used. Many institutional forms that generate actual transactions involve devices to lower a specific transaction cost that would otherwise be high.

One widely-occurring mechanism for lowering transaction costs is that an organization, such as an NGO or private firm, conducts transactions as the representative of consumers who wish to buy the NCO service that the organization coordinates. The organization carries out the negotiations, contracting, and monitoring. Without this mechanism, each consumer would have to undertake each of these steps, resulting in extremely high transaction costs. In general, monitoring cost is not low because a large number of farmers participate in each scheme and, further, they are spread over a large area. Coordinating organizations delegate or entrust the monitoring work to their local or affiliated organizations to reduce costs. The negotiation and preparation of contracts for the easement arrangements (CS4) are high since these are tailor-made and have permanent legal force. These costs may be reduced by employing a standard contract format and simplifying required procedures. An information cost to make potential consumers aware of the existence of an NCO service, namely an advertisement cost, is high in the case of the market price premium scheme (CS2), because such an arrangement is new to consumers in the UK.

A free-rider exclusion mechanism, whether physical or institutional, also incurs a transaction cost since it protects the right of consumers who are eligible to exclusive enjoyment of a particular service. The relevant cost can be thus categorized as an enforcement cost based on the categorization provided above. As mentioned earlier, no explicit exclusion mechanisms are adopted in any of the cases examined, but it can be assumed that setting up and operationalising the mechanism would hinder NCO provision due to high costs.

Government interventions

In most cases, governments intervene in various ways, either in developing legislative frameworks, providing financial assistance, or providing information. In the cases of the National Trust (CS3), Italian agritourism (CS1), and easement arrangements (CS4), there is legislation that supports the activities or provides basic rules. Legislation recognizing that a certain NCO needs to be conserved provides a strong message that societies support and promote the activities in question in a sustainable manner. The legislation thus assigns property rights to farmers who conserve the NCO. Financial supports from government in the market price premium scheme (CS2) and the Austrian landscape program (CS5) are also a sign from government that it wishes to promote certain activities. Other government interventions include information provision, as observed in the Japanese consumer movement (CS8).

Some cases would be at risk or would not be operational without substantial government financial support under existing programmes that supplement the income of participating farmers. Such financial transfers are not specifically intended to help NGAs, but are, nonetheless, an important financial basis for the scheme as observed in the National Trust (CS3) and the Norwegian case (CS6). Depending on site-specific factors, government interventions may contribute to secure environmental values, such as landscapes or biodiversity. It should be noted also, however, that some government interventions, especially those that encourage production, may actually cause pollution, loss of landscape or biodiversity, and may therefore threaten the environmental or other NCOs that society wishes to be preserved.

Voluntary provisions

Voluntary provisions may be numerous but are mostly on a small scale and not documented. Therefore, their impact on societies is unknown. Voluntary provisions are defined in this study as ones with no private pecuniary transactions between NCO suppliers and consumers. Motivations and institutional settings are key elements to examine. As long as no financial compensation for NCO suppliers is involved, motivation must come from domains relating to social aspects such as altruism or solidarity. As observed in the case of voluntary flood mitigation in Japan (CS8), farmers motivated by community solidarity voluntarily mitigate floods affecting downstream community members although upstream farmers risk some loss in rice yields by their activities. In principle, there are no agreements and obligations, but tacit rules among the farmers govern the entire system and reduce transaction costs. In the voluntary provision case in New Zealand (CS9), landowners, including farmers, voluntarily protect biodiversity and landscape through conservation covenants and related activities so as to protect the local environment when threatened by development pressure and to be proactive rather than just accepting State imposition of regulations, such as zoning. Both a bottom-up approach and a strengthened farming community network are major contributors to the success of this case.

In contrast to formal rules (laws, regulations, etc.), informal rules (norms, customs, etc.) are much more difficult to change (Eggertsson, 1996). In general, voluntary provisions seem to be motivated by such informal rules, and therefore the provision is likely to be sustainable once it is undertaken. However, its establishment could be rather difficult. There seems to be no common criteria to design institutional settings fitting to heterogeneous individual settings.

Assessment of NGAs

The main focus of this section is on efficiency, equity and stability. Governmental provision is sometimes compared with NGAs. The feature of multiple outputs from agriculture also needs to be examined as it is significant to the efficiency and equity discussion. Simultaneous NCO provision by government and non-government sectors, which pose a practical problem in designing policies, are then examined. The concluding section summarizes these assessments and develops some policy implications.

Efficiency

As discussed earlier, there are no clear exclusion mechanisms in most of the cases studies. In economic theory, without such mechanisms, no transactions would take place because consumers can enjoy a benefit of a service without paying. However, transactions actually take place. Ignoring possible free-riders, the price of NCO services is determined on the basis of the cost necessary to produce one unit of good consumed by an additional consumer. The price reflects a level of demand based on consumers' willingness to pay, which leads to equilibrium between supply and demand, resulting in efficient resource allocation. The type of market mechanisms that many NGAs rely on gives clear valuation information and demand and supply are determined through the price of the service and willingness to pay. In contrast, governments may determine preferences for NCO services through surveys, workshops, or hearings. These provide important information, but questions could arise as to how accurately they can estimate aggregate preference.

An essential problem is how to deal with the potential demand of free-riders. It is difficult to determine the existence or significance of free-riders. But if they exist and have preferences similar to those of actual consumers, the service is under-provided, leading to inefficiency. Government provision of the service is an option if free-riders are significant. However, as provision is then free of charge, it would bring about a further welfare loss.

If both government and non-government can provide the same NCO, cost-effectiveness is important in assessing the outcome. The associated costs are largely transaction costs. Conceptually the main advantage of governmental provision is scale. A large number of farmers may participate in a government program. Governments can develop uniform contracts and use networks to enforce and monitor them even in remote areas. However, the more they pursue precision by applying a flexible approach, the more administration costs expand. In this regard, Wiebe *et al.* (1996) stated that non-profit organizations offer flexibility and agility, the ability to mobilize private financial and political support, and the capacity to provide local knowledge and insight. In addition, Hodge (2004) indicated that "Such [non-government] organizations can act entrepreneurially, seeking new products and new methods of achieving conservation goals."

Equity

As discussed in OECD (2001a), equity issues can be considered from two angles, *i.e.* benefit and cost implications. The focus of the former is how benefits of multiple NCOs simultaneously produced are distributed among populations and areas with different levels of income and willingness to pay. With respect to the latter, the cost of NCO provision affects, or is affected by, individual income distribution. It should be noted that equity as discussed here is a domestic issue, but that international distributive implications should not be forgotten. If there is an externality associated with goods internationally traded, internalization of the externality, including NGAs, affect the welfare (income distribution) of both exporting and importing countries. However, this issue is not taken up in the present study.

Benefit implications: NCO services provided through NGAs seem to be biased to consumers with relatively high incomes. Income elasticities of demand vary among NCOs. As income increases, demand for environmental services (such as preserving specific species) seems to increase; while that for food security seems to decrease (OECD, 2001a). Unlike land conservation or food security, the environmental services do not have direct impacts on survival or assets of consumers, but rather are consumed for their leisure or in order to conserve resources for future generations. Thus, the rich group tends to value the service more, but in most cases anyone can enjoy them as there is usually no exclusion mechanism, which implies that this is not an issue.

Private transactions of NCOs tend to take place in specific areas having high tourism or biological values or with respect to farmlands on the verge of disappearance due to development. The areas having high value tourism or biological resources attract more demand, thereby are likely to lead to higher revenue for suppliers. Other areas where ordinary environment are involved but which could nevertheless produce more NCOs, tend to be left out.

Cost implications: An equitable solution is that payment should correspond to the service received. With regard to NGAs, generally some pay and others free-ride although they all enjoy almost equal benefits. Inequity exists in this respect, but there seems to be no explicit complaints concerning the payments made voluntarily, and free-riders remain silent. The cost aspect does not therefore seem to be an issue, as long as the payers are willing to tolerate free riders.

Stability

The stability of NGAs is hard to assess since there is no objective standard to assess future uncertainties. Nonetheless, this section examines financial stability and other aspects that are relevant to stable NCO provision under NGAs. In addition to internal factors, external factors, such as market forces and government policies, also affect stability and may be determinant. Lastly, the comparison with governmental provision is important but difficult.

Financial mechanisms, especially income, are crucial for stability. Revenue sources differ across the case studies: donations, sales of food commodities, tax on tourists, tax reductions, or lump sum payments. The first three are paid by individuals, which means that each scheme needs to keep prospective consumers committed on a recurring basis. The last two are paid just once or within a short period. This type is more secure since the relatively large payment is made at an early stage in the operation of the scheme.

Aside from financial arrangements, there are other elements that are important in securing stability of NCO transactions. Arrangement that are legally binding are likely to be very stable, as can be seen in the case of easement arrangements, in which the farmers' commitment to continue agriculture is legally binding; an exit from the commitment requires a court decision, *i.e.* a high transaction cost. The National Trust has legal status, a long history of operation and a very large membership. These factors together assure the sustainability of this mechanism. The landscape conservation program in Austria is another example of a stable system. The programme is operated by a community because its landscape, consisting of a specific type of architecture and surrounding farmlands, is a vital resource for the tourism industry. This mutual reliance between agriculture and tourism reinforces financial viability of both sectors, thereby creating a more secure system.

Market conditions and government policies, especially agricultural reform, play a crucial role in sustaining NGAs. As can be seen in the case of the National Trust (CS3), the continuation of agriculture for some of its tenant farmers is jeopardized because of recent low commodity prices and resultant low profitability. It could therefore be said that stable NGAs of the kind associated with the continuation of farming require a healthy market environment.

A simple comparison with public provision is impossible, but government provision may preserve greater continuity once the initial provision level is determined (OECD, 2001a). The decision to provide a particular NCO is made through a political process that reflects broad demand from the public. As long as the demand does not significantly change, the provision will most likely continue. However, in reality, policy changes resulting from budgetary pressure recently observed in many countries, may interrupt the continuity of public provision. In addition, it is unknown how accurately the political process captures demand because it is not completely transparent and there are technical limits to the capacity to estimate demand. These elements could hinder stable public provision.

Multiple NCOs

Multiple NCOs may be linked simultaneously to the same agricultural production process. Provision of one NCO through an NGA may therefore also contribute to the production of other NCOs'. Some demand for these other NCOs is, therefore, met without any further actions being taken. The extent to which this occurs will differ according to the type of agricultural production in question and the links of the multiple NCOs to production and to each other. Equity may be improved if NCOs such as food security[5] that are valued more by those with low income are produced. But no evidence was found in the case studies.

Governmental and non-governmental provision

NGAs and government agri-environmental payments sometimes have common objectives and outcomes. If the government payments precede NGAs, the latter could be crowded out. Alternatively they may co-exist by physical demarcation of the areas or activities subject to governmental provision. No evidence was found for crowding-out, but private entities perhaps easily avoid duplicate investment. The crowding-out effect could mean fewer innovative private initiatives. Meanwhile, the replacement of governmental by private provision could lead to Pareto improvement because potential willingness to pay for a specific service replaces broad demand and taxation, which may include populations that are actually indifferent to the NCO services being provided. In practice, private entities are likely to want to continue to free-ride the benefits generated by public provision and to invest where government does not provide NCO services. Demarcation could also occur. By this, landowners or farmers perhaps wish to maximise total revenue. Pareto improvement may occur if government steps out and private provision takes over the activities previously carried out. However, this depends on site-specific factors. For example, if private provision eliminates free-riders, Pareto improvement may not occur because the excluded demand would reduce social surplus.

When private precedes public provision on the other hand, public provision either does not take place or co-exists. The former case is not an issue. The latter case was observed. To maximise financial resources, a private entity undertaking private NCO provision uses government agri-environmental payment programs, to supplement its own resources. This has helped to increase conservation capacity. This is a rational option for the private entity, but in a sense involves free-riding on public provision.

Conclusions and assessment concerning NCOs

This study does not offer policy recommendations but describes how actual NGAs have developed and draws some policy implications that may help policy makers.

- NGAs can generate equilibrium NCO provision and result in efficient resource allocation if free-riders are not taken into account in the analysis. The efficiency of government attempts to achieve the same outcomes is unknown.

- Most NGAs currently developed provide environmental goods. There are two types of such goods: those with high values (because they are rare for example) and those, generally more widespread, that have lower values. Because NGAs develop in order to pursue specific objectives and the interests of specific groups, the physical areas covered are naturally limited.

- With respect to stability, a general assessment of NGAs is impossible, but various mechanisms in each scheme seem to favour the stable provision of NCOs, including secured revenues, legal bindings, and mutual financial dependence between agriculture and tourism industries. In some cases, a mixed approach of collaboration between government and non-government seems to contribute to stability. The stability of governmental provision may be threatened by fiscal problems and political processes. But the comparison between the two modes of provision is difficult.

- Private transactions tend to take place in specific areas having high tourism or biological values or farmlands on the verge of disappearance due to development. This type of service provision seems to be biased towards high income earners. This bias could nonetheless be mitigated by the fact that agriculture produces multiple NCOs, some of which could be beneficial to poorer group that have not actually paid for them. Moreover, if there are no clear exclusion mechanisms everyone enjoys the benefits of NGA provision.

- In most cases, governments intervene in various ways, either in developing legislative frameworks that assign property rights, providing financial assistance, or providing information. In a few cases, government payments contribute to the solid financial basis of NGAs.

- NGAs and governmental agri-environmental payments sometimes have common objectives and outcomes. In this case, private provision could be crowded out or some demarcation between the two modes of provision is necessary. Pareto improvement could occur if private provision replaces public provision, because demand is more accurately captured.

Most countries have, perhaps unintentionally, chosen a mix of public and private NCO provision. There are two options for governments to deal with non-governmental initiatives: encouragement (including shift from governmental to nongovernmental provisions) or *laissez-faire,* depending on the comparative evaluation of non-governmental and governmental provisions.

Based on the observed case studies of NCO provision, it would seem that government alone cannot interpret society's demand for non-market goods. On the other hand, private provision captures demand or at least part of it, but does not seem to solve the problem of free-riders. The optimal point for NCO provision may not therefore be one or the other, but in between. Governments could play a role in optimizing nation-wide NCO provision composed of both governmental and non-governmental elements.

In almost all OECD countries, private actions, while still limited, are found in such areas as agritourism where farmers privately sell NCOs. If public interest and demand for NCOs grows in a society, various private actions to capture that interest/demand are likely to emerge, as the UK has experienced. In this case, whichever policy toward NGAs is chosen by the government, the involvement of actors outside the public domain, including farmers, industries and NGOs who are keen to initiate NCO transactions, is important and communication with them allows for improved strategies as well as stronger social acceptance with regard to government policies (Brouwer, 2004). Several of the case studies presented in the annex provide useful examples for future actions, although they cannot always be transferred to other countries. This is particularly the case with social factors, such as a sense of altruism, that are often determinant in the development of a scheme, and which may be specific to particular cultures. However, the mechanisms can be examined, and those which are innovative, or less affected by social and cultural factors can be applied to other societies.

Assessment of non-governmental approaches: negative effects (NEs)

This project examined three types of private initiatives covering agricultural NEs: direct transactions, farmer-led voluntary approaches (FLVAs), and tradeable permits. Among the case studies, a direct transaction case (a mineral water bottler) resolved a dispute concerning underground water polluted by intensive agriculture. Five sub-case studies mainly focused on soil and water problems, cover the issue of FLVAs. The victims of land degradation are often farmers who are also polluters, and this has become a key factor for FLVAs. Two tradeable permit cases deal with water pollution and wetland eco-systems, respectively. Additional information from the available literature is used to supplement the information from the case studies.

OECD (2001b) provided agri-environmental policy evaluation criteria. Six elements in the criteria (economic efficiency, cost-effectiveness, flexibility, enforceability, transparency/fairness/equity, and policy compatibility) are required for a comprehensive evaluation. The major focus of this study is on environmental effectiveness and economic efficiency of NGAs. The former allows for an assessment of the extent to which the policy meets its intended environmental objectives, while the latter allows for an assessment of the extent to which policy can achieve its stated objective at minimum cost when both compliance costs across all affected parties and transaction costs are included (OECD, 1997).

Issues relating to direct transactions

This study examines one direct transaction case (CS10). To protect water quality from a mineral water spring and to avoid falling short of European regulation, limiting the maximum level of nitrate for mineral water, a bottling firm (the pollutee) negotiated changes in agricultural practices in order to reduce nitrate pollutants emitted from livestock manure or fertilizer residues. The main findings are:

- **Private sector profitability**: Since a high level of investment is required for the purchase of property rights (land acquisition and practice changes) from the polluters, this approach is available only for agents whose businesses are highly profitable.

- **Clear and easy identification of stakeholders and property rights**: In contrast to more general agri-environmental issues, the number of farmers who affect the bottler's spring is limited, and the victim is a single firm. The greater the number of stakeholders involved, the more complex the negotiation process becomes and the more financial resources for both compensation and transaction costs are needed. Property rights clearly delineating production choices and the right transfers to the right agent are key factors for a successful solution.

- **Strong involvement by researchers**: No direct government intervention was necessary to define, implement or enforce this private agreement. Scientists from public research agencies played a strong role in defining precisely the property rights relevant to achieve the desired water quality and in improving mutual comprehension between the *a priori* divergent and asymmetric interests of the two parties.

For the agri-environmental sector in general, polluters are supposed to bear the expense of implementing pollution prevention measures or paying for damage caused by pollution beyond a certain reference level (OECD, 2001b). The mineral bottler wanted farmers to reach a higher environment target than the reference level. To this end, the property rights in so far as production choices were concerned were transferred to the bottlers from the farmers, who, of course, received payment. This is a typical case of a Coasian solution, in which the agent who pays under a decentralized bargaining arrangement is determined by the initial assignment of property rights, unlike Pigovian rules in which polluters always pay for the right to pollute (Garvie, 1999). In general, Coasian solutions are not well-suited to environmental problems because the property rights are poorly identified, substantive coordination among large numbers of polluters and victims is required, and there is imperfect information on cost and benefits of pollution abatement (*Ibid*). However, in the above case, all of these institutional and other hurdles were cleared.

Direct transactions can bring about Pareto-optimal solutions through decentralized negotiations, but such cases are successful only under limited conditions. As indicated in the findings, high profitability for pollutees and the clear identification of both actors are the minimum conditions necessary for this type of solution. So far, other mineral water bottlers are considering or have actually taken the same approach.

Issues relating to farmer-led voluntary approaches

FLVAs are represented by landcare-type activities in which government intervention is limited to technical assistance and/or minimum financial support. Four sub-case studies in four countries (CS11), Australia, New Zealand, Canada, and the Netherlands, were conducted in a 1998 OECD study in which voluntary farm community groups were formed to manage, mostly local, environmental problems. Their main concern was initially soil related problems and later expanded to water quality under the influence of the regulating authorities. Other issues, such as pest and weed management or landscape were also included in some cases. The activities common to each of the four sub-cases were: physical works such as planting trees and earth works, the creation and sharing of new knowledge, technical support from external institutions, and the preparation of comprehensive farm planning taking into account the social, environmental, and economic features of farms. Another New Zealand case (CS12) initiated by the wine industry as a marketing strategy, is developing sustainable wine grape production, in which farmers voluntarily participate in mitigating negative effects such as nutrient leaching and soil degradation. The program developed a science-based scorecard approach in which various factors for better environmental management are monitored and assessed. Benefits for the farmers include the use of the logo which is an asset in marketing as well as the environmental improvements themselves. The major findings from these case studies were the following.

- In CS11, the impact on the environment and the sustainable use of resources could not be assessed as data were not available, but some positive effects, such as the reduction of wind erosion, were perceived by participating farmers. In contrast, environmental targets in CS12 were attained by setting the maximum application rate of fertilizers and pesticides together with external auditing.

- Awareness of agri-environmental issues and development of relevant knowledge and skills was the most tangible accomplishment and benefit.

- The provision of a social focus for rural communities helped farmers to identify the root causes of the environmental issues, not just the symptoms.

- The collaboration with research institutes helped farmers to obtain new technologies and enabled researchers to set appropriate research priorities. The collaboration with industry also helped to produce commodities reflecting market preferences.

- A bottom-up approach was recognized as an effective instrument to develop policy.

- Transaction cost in particular in CS12 seemed to be low in comparison with regulatory measures, since the program is self-funding and uses an electronic data and information system.

The cost effectiveness of FLVAs is difficult to assess because of the lack of data and because voluntary community groups are very heterogeneous. Many farmers who completed their farm plans began to monitor their performance against economic, physical and biological conditions, as well as conduct evaluation studies. Voluntary programs (not just FLVAs), nonetheless, seem to be generally more efficient than compulsory pollution abatement (Wu and Babcock, 1999).

FLVAs adopt more or less bottom-up approaches, contrary to the top-down approach used in CAC regimes. Bottom-up approaches are suitable when the solutions sought need to be applied to heterogeneous farm conditions because they allow greater flexibility and freedom to find cost-effective solutions that are tailored to specific conditions (Segerson and Miceli, 1998). This institutional setting generates innovation because of collaborative research with institutes. Farmers seem to have more confidence that working as a group they would be more likely to achieve satisfactory, locally accepted results than if they were to wait for government solutions.

Good institutional settings build in mechanisms in which all stakeholders are motivated to participate actively to achieve shared goals. For FLVAs, raising farmers' awareness of the necessity to reduce negative effects of farming is a core target. As reported, the first tangible outcome is increasing farmer awareness, which in turn works as an engine to sustain activities. The improvement of soil quality is a good target as it is directly linked to farm yield in the long run, and thus it is something that most farmers are easily motivated to undertake. In contrast, it is difficult for a farmer to spontaneously undertake the improvement of water quality outside a regulatory framework or without some kind of incentive, in view of their role as polluters. When greater emphasis in a programme is given to water than to soil, governments may have to pay a relatively larger financial incentive under voluntary approaches. In addition to farmers, industries, research institutions and non-farmer community members often take part in FLVA programmes. Research institutions, public or private, are informed of research priorities in return for the provision of relevant technologies; environmental issues impacting on an entire community are sometimes dealt with.

Issues relating to market-based approaches: tradeable permits/credits/quotas

The use of tradeable permits/credits/quotas is growing for both domestic and global environmental problems. For the agri-environmental sector, they are used to tackle problems ranging from water nitrate pollution, to livestock waste pollution, the allocation of irrigation water, and wetland development rights. Thus, market based approaches may be employed, not only for pollution problems but also for the allocation of natural resources. This study examined two actual cases: tradeable manure/animal quota scheme in the Netherlands (CS13) and the wetland mitigation banking scheme in the US (CS14), which are respectively classified as cap-and-trade and baseline-and-credit schemes (OECD, 2004). The first case restricts the total amount of pollution, and creates quotas that can be traded, while the latter issues credits for resource users to ensure that the total amount of resources respects the baseline. The first approach deals with environmental "bads" and the second approach aims to conserve "goods". The resulting differences are considerable and any common conclusion is difficult. In addition, compared to other sectors, the practical application of market mechanisms in the agricultural sector is relatively recent and examples are limited in number and in countries.

The mechanism for the tradeable permits/credits/quotas scheme largely consists of two basic elements: the setting of physical restrictions in the form of rights, permits or obligations on agents; and the permission to transfer, under certain conditions specified by a regulatory authority, these obligations and rights between agents (Kampas and White, 2003). It is generally recognized that this mechanism can achieve the same environmental quality standard as CAC schemes, but at a lower cost since agents meet a constraint set by the authority and their permits/rights are reallocated to their highest value use through market exchange. Cost-effective technological innovation can be encouraged by this process. Randall and Taylor (2000) suggest that incentive-based policies, including tradeable permit schemes, may achieve considerably lower regulatory costs than can CAC policies. The main findings and issues from the case studies are the following.

- *Environmental effectiveness*: The Dutch quota system (CS13) can potentially prevent further increases in environmental pressure by imposing a farm cap on the quantity of manure production or animal numbers. However, changes in the distribution of livestock resulting from the introduction of trading in manure/animal production quotas have not occurred. It seems that trading has been limited by other policy measures, such as milk quotas, or local community resistance, although redistribution may have been economically viable. For the mitigation banking scheme (CS14), from 1993 to 2000, about 24 000 acres of wetlands were allowed to disappear, and 42 000 acres were required as compensatory mitigation on an annual basis. However, the literature on compensatory mitigation suggests that required mitigation projects are not often undertaken or fail to meet permit conditions, and their magnitude is not known from current data (National Academy of Science, 2003). Difficulties, such as the technical complexities involved in restoring or creating a dynamic wetland system or institutional problems related to uncertainties as mentioned below, could explain these failures.

- *Facilitation of trading*: The permit market is governed by trading rules set by regulators to achieve specific policy targets. The detailed rules determine the level of trading. In the Dutch case (CS13), a manure management policy, aiming to reduce a regional imbalance for the pig and poultry sectors while capping the total manure produced nationally, imposes a number of trading rules and regulations, thereby limiting market participants and reducing market size. Likewise, various rules are enforced under the mitigation banking scheme (CS14) targeting "no net loss" of wetlands. These requirements need administrative approval and therefore raise transaction costs. In general, the establishment of rules requires regulators to monitor participant compliance and thus works as a disincentive. The greater the number of rules, the higher the transaction costs for either regulators or market participants, or both, possibly leading to an inactive market.

- *Initial allocation*: The initial allocation of permits/credits/quotas is considered as the most controversial aspect in the design of a tradeable scheme (Tientenberg, 2004) because trading involves the transfer of revenue, *i.e.* income effects (Kampas and White, 2003). An initial allocation based on historical emissions minimizes political controversy as existing polluters are no worse off than before (Tientenberg, 2004). It is, nonetheless, cautiously suggested that polluters are tempted to increase the current pollution level in order to get larger future permit allocations (*Ibid*), as demonstrated in the Dutch case (CS13). The initial over-allocation in this case study restricted trading that would otherwise have been triggered by improved management or technological heterogeneity. Nonetheless, trading mostly occurs between farmers who wish to increase their farm size and those quitting the business.

- *Uncertainties*: A common and inherent impediment to all trading programs is uncertainty concerning, for example, future total discharge limits, the enforcement of existing policies, and the cost and effectiveness of control technologies (Randall and Taylor, 2000). Technical uncertainties inhibit prospective market participants from using the mitigation banking scheme (CS14). In the Dutch case (CS13), uncertainties regarding the continuation of future quota systems and the introduction of further constraints had a negative effect on the operation of the quota market.

- *Nonpoint source pollution*: The application of the scheme to nonpoint source pollution is a relatively new challenge for the agri-environment sector. Recently in the US, point-nonpoint source trading has been developed in which water quality trading is pursued. But this market differs from the ordinary trading permit markets; that is, non-point sources are always sellers (Taylor *et al.*, 2003). The main issue here is an asymmetric information problem inherent in nonpoint source abatement, which is specifically attributed to the inability to observe individual farmers' abatement actions (technologies) and their costs, and which results in moral hazard (*ibid*).

The primary objective for the introduction of the tradeable permit scheme is to reduce welfare loss compared to what would be generated from a non-trading CAC regime. In the absence of trading, its environmental effectiveness is the same as a CAC regime as it caps total pollution. The occurrence of trading is a factor determining whether the scheme is effective or not. On the other hand, the definition of the baseline and the establishment of administrative procedures by which credits are created lead to tremendous environmental effectiveness for the baseline-and-credit scheme (OECD, 2004). Trading of credits is also important for efficient resource use or environmental effectiveness depending on how resources are conserved under a credit trading scheme.

Tradeable permit schemes generally involve trading rules that are necessary to govern the trading regime. However, the rules seem to cause high transaction costs and reduce trading levels (Stavins, 1995) as can be observed in the case studies. Further to this aspect, initial permit allocations, uncertainties and consistency with other agricultural production policies are key elements in the design and implementation of the scheme. Lastly, trading programs for non-point pollution are a new challenge and an alternative to conventional agri-environmental policy tools. As in the Dutch case (CS13), the introduction of observable parameters — the amount of P_2O_5 and the number of animals producing it — into a scheme to deal with non-point source pollution enables makes trading viable which it would not be if it was dependent on actual measurement of emissions from different farms. Taylor *et al.* (2003) propose a collective performance-based contract for the nonpoint source pollution trade, which concentrates on the observable level of aggregate non-point source abatement produced by multiple agents. This, however, needs to overcome the moral hazard problem as individual abatement is not observable.

Conclusions and general assessment concerning NEs

Given the conditions under which direct transactions occur, pure non-governmental solutions where polluters and victims determine optimal solutions through negotiations cannot prevail in most agri-environmental settings. FLVAs such as Landcare type projects are close to pure non-governmental solutions with decentralized negotiations and activities in a community, in which polluters and victims, seek their own solutions for on-and-off-farm resource degradation problems, especially for soil and water. In principle, these solutions do not require substantial government intervention, but they do often need technical and administrative support as a catalyst. Tradeable permit schemes, meanwhile, rely on pollution caps or resource use restriction set by governments. However, the decentralized negotiations among polluters leading to trading reduce welfare loss compared to pure governmental solutions.

Clear identification of property rights is a key factor for success, especially for non-point source pollution. Successful direct transactions are an example, but tradeable permit schemes also include this mechanism in the form of pollution permits, credits, or quotas. For FLVAs, instead of the identification of individual property rights, the scheme urges increased awareness that public resources should be properly conserved within a community, and that on-farm resources can be better conserved by sharing knowledge.

The success or failure of the approaches examined in this study is, in large part, determined by the strength of compliance incentives provided by institutional settings (Garvie, 1999). In this sense, such institutional settings are an essential mechanism for any NGAs. All three approaches incorporate some form of financial incentive. Such incentives are generated from: the improved profitability to be gained from an improved environment; the awareness of the necessity for combined environmental and financial sustainability; and the revenue gained from differences in technologies for pollution abatement or efficient resource use.

Conclusions

A total of eight types of NGAs have been examined in this study. Strengths and weaknesses have been identified with respect to each type, which makes it difficult to unreservedly recommend specific solutions. Nonetheless it should be noted that the various innovations in NGAs that have been observed seem to provide significant potential to cope with externality problems for example, the market price premium scheme for NCOs and tradeable permit programs, and voluntary approaches for NEs. It is unknown at this stage whether these new measures will be successful, but early results are promising. The market premium scheme allows NCOs to be transacted in an existing consumer market and, as the premium is relatively small, a wide population can access the NCO service. This is in contrast with some other types of NGAs that are largely captured by a relatively small interest group. Meanwhile, the agricultural non-point source pollution problem, which remains one of the most difficult policy challenges (Peterson and Boisvert, 2001), is tackled by polluters themselves under tradeable permit programs or voluntary approaches that provide them with greater flexibility to choose the most appropriate technologies and continue to pursue innovations.

Property rights are key elements in dealing with externality problems for both NCOs and NEs. Under well-defined property rights, NCO producers are identified, the NCO is valued, priced, and then transacted; for NEs, who pays for what and on what terms is also determined. As stated in the introduction, an important government role in the formation of NGAs is to define property rights. Some of the most successful cases of NGAs for both NCO and NEs provision examined in this study depended on government intervention to define property rights. The process in which property rights are assigned to a particular party may sometimes require scientific expertise as agriculture involves complex and dynamic eco-systems; enforcement is also important in avoiding free-rider problems; and finally definition of property rights is a key factor in valuation. Despite the limited number of cases examined it is clear that government's role in defining property rights is crucial in the development of NGAs for both NCOs and NEs.

Notes

1. One case for this type of transaction is reported in the present study.

2. Indepth analysis on voluntary approaches for the sectors other than agriculture is contained in OECD (2003b).

3. This is exemplified by a private provision of bridge construction (Stiglitz, 1988). If the private company charges a toll, some potential beneficiaries would be excluded although the cost of providing an extra trip is almost zero. This, therefore, generates welfare loss.

4. There is growing evidence that good economic performance is compatible with good environmental performance. For example, the U.S. Economic Research Service found that U.S. corn producers who use crop residue management (CRM) to minimize damage from agricultural runoff enjoy a clear economic edge over non-CRM corn producers (Hopkins and Johansson, 2004).

5. For food security, the degree of jointness with domestic food production is controversial.

References

Barthelemy, D. and M. Nieddu (2004), "Multifunctionality as a Concept of Duality in Economics: An Institutionalist Approach", the Proceedings of 90th European Association of Agricultural Economics Seminar, http://merlin.lusignan.inra.fr:8080/eaae/website

Brouwer, F. (2004) "Introduction", in F. Brouwer (ed.), *Sustaining Agriculture and the Rural Environment: Governance, Policy and Multifunctionality*, Edward Elgar, Cheltenham, UK.

Buller, H. and C. Morris (2004), "Growing Goods: the Market, the State, and Sustainable Food Production", *Environment and Planning*, Vol. 36, pp. 1065-1084.

Coase, R. H. (1960),"The Problem of Social Cost", *Journal of Law and Economics*, Vol. 3, pp. 1-44.

Cornes, R. and T. Sandler (1996), *"The Theory of Externalities, Public Goods, and Club Goods"*, Second Edition, Cambridge University Press.

Department of Agriculture, Fisheries and Forestry of Australian Government (2003), *"Review of the National Landcare Program"*, Report.

Eggertsson, Thráinn (1996), "A Note on the Economics of Institutions," in *Empirical Studies in Institutional Change*, edited by L. T. Alston, T. Eggertsson, and D. C. North, Cambridge University Press, New York and Cambridge.

Furubotn, E.G. and R. Richter (1998), *"Institution and Economic Theory"*, The University of Michigan Press, USA.

Garvie, D.A. (1999), "Self-Regulation of Pollution: The Role of Market Structure and Consumer Information", in B. Bortolotti and G. Fiorentini (ed.), *Organized Interests and Self-Regulation: An Economic Approach*, pp. 206-235, Oxford University Press.

Heimlich, R.E. and R. Claassen (1998), "Agricultural Conservation Policy at a Crossroads", *Agricultural and Resource Economics Review*, April 1998, pp. 95-107.

Hodge, I. (2004), "Methodology and Action: Economic Rationales and Agri-environmental Policy Choices", in F. Brouwer (ed.), *Sustaining Agriculture and the Rural Environment: Governance, Policy and Multifunctionality*, Edward Elgar, Cheltenham, UK.

Hopkins, J. and R. Johansson (2004), "Beyond Environmental Compliance; Stewardship as Good Business", *Amber Waves*, Volume 2, Issue 2, Economic Research Service of the US, www.ers.usda.gov/amberwaves/ accessed December 2004.

Kampas, A. and B. White (2003), "Selecting Permit Allocation Rules for Agricultural Pollution Control", *Ecological Economics*, Vol. 47, pp 135-147.

Lyon, T.P. and J.W. Maxwell (2003), "Self-regulation, Taxation and Public Voluntary Environmental Agreements", *Journal of Public Economics*, Vol. 87, pp 1453-1486.

National Academy of Science (2003), *"Compensating for Wetland Losses under the Clean Water Act"*, Washington, US.

OECD (1997), *"Evaluating Economic Instruments for Environmental Policy"*, Paris.

OECD (1998), *"Co-operative Approaches to Sustainable Agriculture"*, Paris.

OECD (2001a), *"Multifunctionality: Towards an analytical framework"*, Paris.

OECD (2001b),*"Improving the Environmental Performance of Agriculture"*, Paris.

OECD (2001c), *"Domestic Transferable Permits for Environmental Management: Design and Implementation"*, Paris.

OECD (2003a), *"Multifunctionality: Policy Implication"*, Paris.

OECD (2003b), *"Voluntary Approaches in Environmental Policy: Effectiveness, Efficiency and Usage in Policy Mixes"*, Paris.

OECD (2004), *"Tradeable Permits: Policy Evaluation, Design and Reform"*, Paris.

Peterson, J.M. and R.N. Boisvert (2001), "Control of Nonpoint Source Pollution through Voluntary Incentive-Based Policies: An Application to Nitrate Contamination in New York", *Agriculture and Resource Economics Review*, Vol. 30/2, pp. 127-138.

Randall, A. and M.A. Taylor (2000), "Incentive-Based Solutions to Agricultural Environmental Problems: Recent Developments in Theory and Practice", *Journal of Agricultural and Applied Economics*, Vol. 32(2), pp. 221-234.

Segerson, K. and T. Miceli (1998), "Voluntary Environmental Agreements: Good or Bad News for Environmental Protection?", *Journal of Environmental Economics and Management*, Vol. 36, pp. 109-130.

Stavins, R.N. (1995), "Transaction Costs and Tradeable Permits", *Journal of Environmental Economics and Management*, Vol. 29, pp. 133-148.

Stiglitz, J. E. (1988), *"Economics of the Public Sector"*, Second edition, W.W. Norton & Company, New York.

Taylor, M.A., *et al.* (2003), "A Collective Performance-based Contract for Point-Nonpoint Source Pollution Trading", Selected paper presented at 2003 American Agricultural Economics Association Annual Meeting in Montreal.

Tietenberg, T. (2004), "The Evolution of Emissions Trading: Theoretical Foundations and Design Considerations", http://www.colby.edu/personal/thtienten/ accessed January 2004.

Wiebe, K. *et al.* (1996), *"Partial Interests in Land: Policy Tools for Resource Use and Conservation"*, Agricultural Economic Report No. 744, Economic Research Service, US Department of Agriculture, Washington D.C.

Wu, J. and B.A. Babcock (1999), "The Relative Efficiency of Voluntary vs Mandatory Environmental Regulations", *Journal of Environmental Economics and Management*, Vol. 38, pp. 158-175.

Part II.

Case Studies

1. Agritourism in Italy (Type P1)[1]

Abstract and classification

Agri-tourism involves transactions of NCOs between individual farmers and consumers, and the transactions do not involve any agricultural commodities so that this case study can be categorised in Type P1 in this project.

Background

Agritourism activity has expanded constantly in Italy since its development in the 1980s. The number of farms offering some kind of tourist services has doubled to 12 500 units in 2003, with a turnover of 750 million euros (Agriturist, 2003). Many farms offer accommodation (10 000 units), restaurants (7 500 units), camping facilities (930 units), and horse riding (1 520 units). A recent, but expanding, activity is the development of itineraries that seek to introduce agriculture, rural activities and traditions to families and, in more specialised cases, to schools.

Agritourism farms in Italy are spread over the country (2% of all farms carry out agritourism activities), but are most prevalent in the northern and central regions, particularly in Tuscany (20%). Only in the last decade has there been an expansion of agritourism farms to the southern regions, particularly in Puglia, located in south-east Italy, where the share of agritourism is the same as the national average.

The peninsula of Capo di Leuca is located in Puglia. The agricultural sector is characterised by fragmentation of the farms (1.8 hectares on average) and low levels of production and diversification. In brief, the area is representative of Objective 1 zone,[2] that is, characterised by under-development, such as a high unemployment rate and immigration (Sivini, 2001). Agritourism has developed at a fast rate and currently plays a vital role in preserving the rural way of life, enabling farmers to diversify their activities while contributing to the preservation of the area's historical and cultural heritage.

One of the most typical of these historical structures is the so-called *Masserie*, an agricultural production centre made up of large agricultural estates typical of the late Middle Ages; they represent an interesting cultural crossroad of local history, economics, culture, and agriculture. Farmers currently own several of the *Masserie* which they have restored and use for agritourism purposes. This case study focuses on the farm Masseria Marcurano.

Non-commodity outputs addressed and the mechanism used

The Masseria Marcurano farm consists of seven hectares and cultivates olives, cereals and vegetables. The agritourism activity was initiated in 1995 and the services offered are accommodation, restoration facilities, direct sale of farm products, and two itineraries. One of the itineraries combines cultural heritage with landscape, and the other is of a more naturalistic type; both contribute to the visitor's knowledge of the area's cultural heritage while enjoying its natural resources. The farmer, an artist originally from the north-east Italy, personally restored the Masseria building where accommodation is offered.

The historic itinerary consists of a trekking route lasting four hours during which visitors are introduced to the history of the local agriculture system and its techniques. Visitors visit a sixteenth century peasant settlement made by *Pagghiara*, a characteristic local structure made only of stones and known as *Trulli*. Another point of interest is the Canal of Fano due to the presence of ancient caves where Basilian monks lived during the Iconoclastic War (between the seventh and eleventh centuries). The monks introduced into the area the olive growing and its techniques imported from the Orient.

The second itinerary is geared more towards nature and leisure and consists of a boat trip along the Adriatic and Ionic coasts of the peninsula, visits to some of the numerous coast caves, and sea baths.

The NCOs in this case are landscape and cultural heritage. Landscape is composed both of olive trees present on the farm and its surroundings, combined with typical rural constructions. In the case of olive groves, there is a direct link with agricultural production: while producing olive oil the farmer cares for the groves undergrowth and implements controls against pests and weeds that contributes to the maintenance of the landscape. In the case of the rural buildings the linkage exists — the Masseria would be abandoned if there was no farm activity— but, in theory, other institutions or individuals could manage and maintain them. Under the Italian and regional legislation, however, farmers can enrol in agritourism activities only if they continue to carry out agricultural activities. Consequently, the law imposes the linkage. Although both NCOs could be obtained by having visitors rent rooms in the Masseria as well as enjoying the landscape, these NCO are enhanced during the trekking itinerary. In fact, the farmer acts as a guide and introduces visitors to the local rural history and olive growing techniques from the Orient.

During the low season, accommodation facilities are provided on a per-day rate of EUR 35 per person, with breakfast included. For the months of July and August accommodation is provided on a per week basis only starting every Saturday. The rate per person varies between EUR 290 and EUR 330 for one week for a double room, depending on the period with the higher rates charged during August. The rate includes breakfast plus the two trips mentioned above, their incidence is estimated as 10-15% of the total weekly fee.

There is neither a differentiation in prices between rooms nor a great difference with respect to prices charged by similar agritourism farms in the area. From the farmer's point of view, the obligation of booking a whole week entails the valuation of the whole set of offered services. Although it is known that some visitors prefer a trekking itinerary to a swimming pool and that others are interested in knowing the Trulli origin, it is not clear how much visitors would be willing to pay for these services. So, NCOs are included in prices in a limited manner.

Institutional settings focusing on governmental policies

At the European level, agritourism is promoted and financed through the European Leader Initiative,[3] whose objective is the promotion of endogenous, integrated and sustainable development in rural areas through Local Action Groups (LAGs). The LAGs are public and private partnerships aimed at revitalising rural areas through an integrated series of interventions that are both economic and social in nature. Among the actions that could be developed by LAGs within the Leader initiative there is one related to rural tourism (action 3).[4] This action provides for the financing of projects aimed at the amelioration of accommodation structures used in agritourism and the creation of networks among farms (INEA, 2000). Support for investment covers 50% of the total value of the investment.

The setting up of the Masseria Marcurana agritourism was facilitated by the actions of the Local Action Group (LAG) *Capo di Leuca* created by the Leader I initiative in 1991. The LAG consists of a partnership between public (municipalities) and private actors (farm trade unions) and it acts as a promoter of integrated local development. The main goal of LAG participation to the Leader I initiative was the realisation of concrete projects aimed at linking small enterprises and the resources of the territory to local and external markets.

In the area of agritourism, the LAG activity consisted particularly in supporting projects aimed at creating agritourism structures. Results were encouraging, with five agritourism projects approved and executed. One of these projects led to the starting up of the Masseria Marcurano agritourism activity. It received EUR 30 000 as investment aid, representing about half of the total investment (Atzeni, 2004).

Other LAG supporting activities that facilitated transactions consisted in the organisation of courses on technical aspects related to tourism activity management and initiatives aimed at promoting local products, history, culture and natural resources. The latter were further developed within the Leader II initiative (1996-2001) through the creation of the "Capo di Leuca Flavour and Knowledge Itinerary". Tourism in general was enhanced through the participation to trans-national tourism co-operation projects Eurovillages and Paralelo 40 in which the Leader I beneficiaries could also participate.

Administrative decentralisation in Italy provides for the possibility for the region to modify the set of instruments available within the Leader Initiatives. As regards the new Leader+ Initiative, given that support for agritourism investments is also included in the mainstream rural development measures, the regional administration has chosen to exclude the possibility to finance interventions in favour of agritourism within the Leader+ Initiative (Atzeni, 2004). In order not to duplicate the source of financing, the current Regional Leader Programme (RLP), that is the "operational" instrument for implementing the Leader+ initiative at regional level, maintains the same general organisational structure and some of instruments of the previous initiatives, but does no longer finances the introduction of new agritourism enterprises or the improvement of existing structures. Nevertheless, agritourism farms can benefit indirectly from the other measures contained in the RLP aimed at enhancing tourism.

At the national level, the agritourism activity is ruled by the Framework Law n 730 of 5 December 1985. This law delegates regions to establish norms and practical criteria to be applied within their territory through regional laws and regulations.[5] Law n 730/85 provides a definition of agritourism that helps to understand its linkage with the agricultural production in the Italian context. Agritourism is defined as "any accommodation facility and receptive activity practised by farmers, their families whereas they play an active role within the family, and any form of farm association through the usage of the farm and its structures, the latter should not be currently used by the farmer and his family for production or personal use. The agritourism activities are linked to the farming activities through a "complementary linkage". The definition has a twofold implication: the agritourism activity cannot exist outside the farm and it cannot prevail over other typical farming activities (INEA, 2001). This responds to the objective of giving an agricultural nature to agritourism and at the same time supports the diversification and modernisation of the agricultural sector. Compared with the legislation applied in other EU countries, the Italian case is peculiar (European Commission, 1990). There is a clear differentiation between agritourism and rural tourism from the point of view of the legislation and actors involved. While agritourism is assimilated within the agricultural sector, rural tourism is assimilated to the tourism sector.

The Agriculture Sector Orientation and Modernisation Law of 15 June 2001 integrates the Framework Agritourism Law. Its aim is to update and widen the definition of agriculture activities and agriculture entrepreneurs[6] to be eventually taken for future support at national level. First, and most important, agritourism is included within the definition agricultural activity, among the "related" agricultural activities (article 1). Secondly, an important innovation is the possibility given to agritourism farms to organise recreational, cultural, educational, horse-riding, and sporting activities outside the boundaries of the farm (article 3) in order to contribute to the knowledge of the territory, including the tasting of local farm products (INEA, 2002).

Finally, although agritourism activity is included among the "related" activities of agriculture, from a fiscal point of view both activities are ruled separately. Income taxes and VAT are ruled as any business activity, with the possibility of applying a simplified regime. This is not meant to be real reduction of taxes but a mere simplification of procedures. The Law n 413 of 30 December 1991 provides for this fiscal regime.

At the regional level, the Regional Law n 34 of 22 May 1985 takes in the Framework Law indications. It defines in detail the agritourism activities, engages the Region of Puglia to finance an annual Agritourism Promoting Programme, establishes a Register of Agritourism Operators, and defines a set of initiatives that could be financed by the Region (comprised investments for restructuring accommodation facilities).

Other measures facilitating transactions

A national consortium (Anagritur), composed of the three national agritourism farm associations (Agriturist, Terra Nostra, Turismo Verde), monitors and co-ordinates the activities promoted by the three single organisations. These associations were founded by the three main Italian farm trade unions for the promotion, support and development of agritourism and the enhancement of the rural world. Among their activities probably the promotion activity by Internet is the most important. They also provide fiscal, law, and economic advisory services.

There are numerous, at least fifty, regional and local web sites, available in several languages, from which visitors can book their holidays. This possibility facilitates enormously the knowledge of the Masseria Marcurano to foreign visitors.

Costs and difficulties encountered in getting the mechanism to work

At the beginning of the 1990s, agritourism activity did not exist in the area due to difficulties for farmers to register as Agritourism Operators of the Region of Puglia. This was due essentially to the complexity of the registration process that included multilevel controls of the farm by *ad hoc* commissions, established at the municipal, provincial and regional levels, prior to the authorisation to register. Long delays were common and there was a risk of losing the investment support due to missed registration. Consequently, an important LAG task during the initial phases was to support farmers' projects making regional and local institutions aware of the importance of issuing farmers with the related agritourism authorisations.

One aspect that still creates concern, but that has not prevented the farmer to start the agritourism business, is the relatively high seasonal concentration of visitors during the months of July and August and the difficulties to hire extra personnel for a short period of time. Finding temporal workers appeared to the agritourism owner as a difficulty that he has resolved in recent years by making arrangements with university students.

Assessment of institutional and other arrangements and conclusions

After almost a decade of operation, the scheme described in this case study could be positively evaluated. Visitors appear satisfied, considering that it is not infrequent that some of them return for a second visit. Prices are considered fair and convenient if compared with fares proposed by similar agritourism farms. Results obtained from this scheme should be transferable to other regions and countries with similar institutional settings.

The role of the Capo di Leuca LAG appears to be of great importance. The initial animation activity that regarded the assistance and promotion of agritourism projects and the public authority sensibilisation was essential. Further work among local actors for the development and the promotion and enhancement of rural heritage contributed to the success of the scheme. The promotion of networks between beneficiaries, like the "Capo di Leuca Flavour and Knowledge Itinerary" has lead to clear advantages and has become autonomous from the LAG (Sivini, 2001). The area is currently well known and its cultural heritage highly valued outside Puglia. The stone constructions (*trulli*) became a symbol of the Region's landscape together with its olive groves. The LAG activity has been recognised as determinant for the introduction of elements of innovation within the territory and for the valorisation of the historical and naturalistic resources. Other activities related to tourism in general were promoted by the LAG through the participation in two transnational cooperation networks and a special project for the restructuring of houses in historic towns in order to be used as tourist hostels.

In conclusion, agritourism in Italy points out the presence of a composite setting of relationships between agricultural production and NCOs. Although it is clear that landscape (olive groves) is provided together with agricultural production, the linkage with agricultural activity is not straightforward, especially in the case of NCOs related to buildings and historic places (cultural heritage) because other institutions or individuals could provide them. The legal setting in Italy (agritourism and reorientation laws) gives agritourism the status of agricultural activity and constrains farmers to continue carrying out agricultural production. This establishes a link with production that could be artificial but has important consequences for the agricultural sector. In this case study it implied a return to cultivation of the land and a requalification of the farmer: he has become "a farm holiday worker" with a more positive image. Other interventions related to tourism enhance the development of the area, but their effects do not influence directly the agriculture sector.

Finally, it emerges from this case a quite common valuation uncertainty on the demand side. Both the olive grove landscape and the stone structures of Puglia have an excellent reputation within the country, and are supposed to be highly valued for their aesthetic appearance and the historic linkage. Nevertheless, it is not clear how much the consumer is willing to pay (in terms of greater agritourism fees for example) to stay in an agritourism structure that conveys those public goods due to the lack of objective criteria for doing so.

Notes

1. This case study was prepared by Beatriz E. Velázquez of the Istituto Nazionale di Economia Agraria (INEA).

2. These are areas, where the gross domestic product (GDP) is below 75% of the EU average, that lag behind in their development. To promote the development of Objective 1 zones is the main priority of the European Union's cohesion policy.

3. Interventions for promoting agritourism activities are also included in the measure "Diversification of agricultural activities" defined within the mainstream structural measures included in the Programming period 2000-2006 and in the Rural Development Programmes and Regional Operational Programmes (INEA, 2000). This is not the case in this study.

4. Actions developed by LAGs are included in a "Rural Innovation Programme". Other than support for rural tourism, other actions include vocational training, technical support for rural development, support for small business, local exploitation and marketing of agricultural, forestry and fishery products, preservation and improvement of environment and living conditions.

5. This reflects the decentralised political and administrative structure of the country where regions have an autonomous role in ruling some aspects of the economy.

6. In order to take into account social and economic changes since 1942, the year when previous legislation defining agricultural activities was issued.

2. Market price premiums for milk in the United Kingdom (Type P2)

Abstract and classification

Milk whose production is linked to the conservation of wildlife is available for purchase in the UK in ordinary retail shops. A price premium has been added to the purchase price, with the amount of that premium being used for conservation work by individual farmers supplying the milk. This case can be categorised in Type P2, where NCOs are traded between individuals and with commodity supply.

Background

Agri-Trade Direct Ltd., a dairy farmers' consortium, suffered from long-term low milk prices and struggled like many other dairy production providers in the UK to increase its production and sale of milk. In parallel, the Wildlife Trust (WT), one of UK's outstanding charities working for the conservation of wildlife, was engaged in restoring wildlife, mainly in countryside. These two groups, seemingly incompatible, jointly created a new brand scheme, Wildcare Dairy Product (WDP) Ltd., in April 2002. This scheme introduced a market price premium mechanism with the objective of giving dairy farmers a financial incentive to conserve and enhance biodiversity on their farmlands by using a market where consumers were willing to pay a premium towards that scheme's conservation work. The responsibilities of WDP are mainly the coordination of players involved (*i.e.* largely dairy farmers and the WT), providing logistical support for the sales, and monitoring the production process and the quality of products. The WT provides technical assistance for the dairy farmers in preparing individual farm conservation plans and the implementation of their own wildlife preservation projects which are supported through the sale of milk.

As this business was started in April 2002, it is at the stage where new markets are explored through a concentration of most of its business resources. As a result, sales have grown steadily, but as of April 2004 the initial capital investment had not yet been recuperated.

Non-commodity outputs addressed and the mechanism used

The WDP brand milk,[1] which includes both organic and non-organic milk, was sold at between GBP 0.54-0.83 per litre in the UK as of February 2004, which is almost the same price level as other brands of milk.[2] The premium is set at 5 pence per litre, 3 pence of which is allocated to dairy farmers supplying the milk and the remaining 2 pence is given to the WT for its conservation projects in the UK. WDP sells the milk to ordinary wholesalers and retailers.

Approximately 300 dairy farmers are currently participating in this scheme. The participating farmers are mandated to give up 10% of their land for wildlife habitat management and prepare a "Whole Farm Wildlife Action Plan" with the assistance of local WT staff and advisers of the Farming Wildlife Advisory Group (FWAG), which is also a registered charity and has on-farm expertise in conservation work. Farmers are instructed in the management of their designated 10% wildlife habitat, *e.g.* how to plant hedges and restoration of woodland and wetlands. Wildlife targeted for preservation includes brown hare, skylark, barn owls, etc.

The WT, a network of 47 independent local wildlife trusts have about 440 thousand members and 2 500 reserves covering about 81 000 hectares, some of which are farmlands. In addition to the management of reserved areas, they provide visitors and students with field education and guides. They give farmers advice or assistance on conservation practices that are compatible with profitable farming inside and outside the trust farmlands. The premiums earned from milk sales are not only used for farmers' conservation work, but also allocated to the WT's own conservation projects.

Without effective conservation measures, the quality and area of wildlife habitats would decline in view of the generally intensive farming practices. The emphasis of the scheme is to change this trend by imposing participating farmers to convert a part of their farmlands to wildlife habitats. This can result in less time to engage in farming practices and perhaps in the reduction of yields.

Institutional setting focusing on governmental policies

The premium being transferred to the WT, set at 2 pence per litre, is exempted from any taxation because it goes to a registered charity, while that portion going to farmers, the remaining 3 pence per litre, is subject to income tax. This is because the latter transaction is regarded as a commercial one regardless of its public benefits and consumers may think that the premium is a donation.

The UK Department of Environment, Food and Rural Affairs provided GBP 53 000 to support the scheme as it recognised that preserving wildlife by using market forces was an innovative approach.

A small number of farmers operating within this scheme have entered governmental agri-environment programs, although there is a specific demarcation between the different projects.

Organisational arrangements

An outstanding organisational feature of this case is that it is a well-designed consortium type of business scheme, in which players involved provide their professional expertise. The main players are WDP, which is responsible for the management, the farmers, the WT, and FWAG. The primary purpose of this scheme is profit-oriented, but the players' interests are roughly divided between the productivity of farming and the conservation of wildlife habitats, which seem to be trade-offs in some aspects. Yet, a balance of the two interests is vital for success. The combined pricing of milk products and conservation work gives players a common incentive to increase sales by providing a share of the sales to the WT. To maintain good relations and a fruitful partnership, meetings are organised on a quarterly basis, in addition to frequent communications by telephone and electronic mailing.

The only expertise that had to be imported from outside professionals to start the business was on-farm expertise for the wildlife conservation while minimising the loss of profitability in farming. It relies on the WT and FWAG. While the WT acts as a partner, FWAG is a contractor providing professional advice and monitoring the farmer's conservation work. Only two new staff were recruited by WDP. As a result, the organisational setting up cost was minimal.

Measures to promote NCO transactions

The initial investment required was large, mostly related to advertising costs. UK consumers do not seem to have any interest in brand names so that generally milk is sold in a simple white bottle with a label indicating a minimum of product information.

Consumers are primarily concerned with price and consequently suppliers compete mainly over price. Value added to the products revolves around whether they are organic or non-organic, and normal or reduced fat milk. Under this situation, the launch of a new brand of milk attempted to introduce a different value in the milk market and needed to change consumers' view about how to purchase milk. More than GBP 300 000 in cash has been spent for advertising in various media. However, with the exception of this particular investment, other costs are minor since no new physical facilities had to be installed. Farmers have to prepare and implement a "Whole Farm Wildlife Action Plan". The necessary expertise is new to most of them, but can be provided by WT and FWAG advisers.

Consumers of this milk product buy the NCO at nearby retail shops or supermarkets. The premium is small, and the transaction process is simple. There are no negotiations or bargaining which eliminates associated transaction costs. Once consumers buy the commodity, they expect the supplier to implement his commitments, although they do not monitor how it is done as monitoring could be costly relative to the premium. However, consumers must be properly informed as to the status of wildlife conservation in order to avoid moral hazard. Many consumers would be satisfied with having the sense that they can participate in wildlife protection without monitoring.

FWAG, an independent professional organisation, is contracted to visit all farms on an annual basis to monitor implementation of the farming plans. The WT verifies FWAG's monitoring report. The WT is a stakeholder of the WDP, but this consortium-type organisation which links various parties informally enables the WT to be relatively independent. Therefore, such checks and balances seem to work properly. Furthermore, the WT is mandated to submit an annual report to the national Department of Environment, Food, and Rural Affairs concerning the receipt of funds.

Information on the milk, including the conservation of wildlife habitat, is provided on labels, the WDP's website, newspapers, television and radio. The last three mass-media venues are also used for advertising by disseminating information on the product and thereby increasing sales through greater awareness. A label provides quality information. The website is a convenient tool to provide the public with a large amount of information as well as for making annual inspection reports available to the general public.

In terms of the price of the premium, two factors can be considered which make this scheme successful: the premium is relatively small and it is fixed, *i.e.* 5 pence per litre. A relatively small premium enables a broader range of consumers to contribute irrespective of their income level. Other non-governmental schemes, such as agri-tourism, environmental trusts, and community supported agriculture, seem to be luxury goods[3] in that people of relatively high income demand more than those earning lower incomes. Secondly, milk can be used for price premium-type transactions more easily than other agricultural products. It is bottled, which leads to a homogeneous quantity, and there is little difference in quality over seasons so that the same price and premium can be maintained. In contrast, vegetable prices change between seasons and places. It would be less clear how much money was transferred to farmers for wildlife conservation due to the commodity's changing price. A predetermined premium is transparent and enhances credibility.

Costs and difficulties encountered in getting the mechanism to work

As stated earlier, the major difficulty for the scheme supplier is to persuade consumers that the milk brand is credible and worth buying. In addition to positive advertising costing about GBP 300 000, the scheme had a key strategy in the coordination with an established environmental organisation, which was already popular and generally accepted UK citizens. This is because this environmental organisation was considered by consumers as a guarantee of the milk's quality. Before establishing a partnership with the WT, negotiations with another charitable organisation had almost been concluded, but collapsed in 1999 due to adverse publicity on BSE. It was thought there was a risk of catching BSE from drinking milk. Trustees became jittery about endorsing a milk brand and eventually the prospective partner pulled out. Two months later, the outbreak of Foot and Mouth disease was a further blow. These two events delayed the launch of the scheme for almost two years. The scheme looked for a new partner and eventually the WT agreed to join. A name brand, of which the impact cannot be quantitatively assessed though, is an important element in facilitating transactions.

Although the farmers have adopted wildlife friendly farming practices, in appearance and quality there are no other differences with ordinary milk. The brand milk is thus colleted from farms and processed in a factory separately from other ordinary milk to avoid mixture and to guarantee its production process. This involves an issue of traceability that entails a certain cost shouldered by producers and/or WDP in this case.

Assessment of institutional and other arrangements and conclusions

The two conflicting objectives of this scheme, *i.e.* profitable production and environmental conservation,[4] are reconciled and pursued simultaneously by using a consumer market where the NCO is valued and consumer are willing to pay for it. Apart from the difficulty in changing the consumer's mind on milk products, the applied mechanism is very sophisticated. This scheme is still at an early stage, so that it is premature to assess its overall success within the UK.

It may be difficult to identify all the beneficiaries for the conservation of wildlife habitat as it has the nature of a non-use value. Thus, the scheme's consumers may not be its only beneficiaries but they valuate the NCO by purchasing the milk. The producers who produce the NCO are all compensated. The pricing of the NCO is unique for the reasons described above. This scheme achieves an efficient NCO production system even though the market is small.

Low income consumers may be less interested in wildlife conservation than higher income ones, but this low cost NCO opens the market to both groups. In this sense, the scheme is equitable over income levels. In so far as regional NCO distribution is concerned, this scheme is less equitable because only a limited number of producers can participate due to limited market and sales, which limits NCOs conservation to a small number of areas. As the market grows, however, the area should expand.

Whether this scheme is stable or not depends on how this business becomes a profitable operation. The cost to maintain the institutional/organisational setting seems to be low and the scheme's two conflicting objectives seem to have so far been successfully reconciled. In comparison with the government agri-environmental programs that have similar approaches and objective, this scheme could generate more stable NCO provision. The participation and remuneration of participating farmers would not be limited in time, which contrasts with government programs which are predetermined (the extension/renewal of the programs can be considered). Of course, in this comparison, the scale, the amount of payment, and the impact on the environment also need to be examined.

Notes

1. The brand milk is named "White and Wild Milk".

2. GBP 1.0 is equivalent to EUR 1.49 as of March 2004.

3. The term of 'luxury good' can be also used as the demand of the good goes up by a greater proportion than income rises (Pearce, 1992).

4. There is growing evidence that good economic performance is compatible with good environmental performance (Hopkins and Johansson, 2004).

3. Conservation Trust: The National Trust in the United Kingdom (Type P3)

Abstract and classification

A trust type provision of NCOs is categorised in Type P3 where NCOs are transacted between farmers and a specific group (club provision) separately from commodity outputs (discrete supply). The National Trust (NT) is a charity independent from government that aims to preserve and protect sites of historic interest and natural beauty in England, Wales and Northern Ireland,[1] which include farmlands.

Background

The first land and natural beauty preservation trust was initiated by the Common Preservation Society in 1865. Since then, a total of over 100 trusts[2] have been launched in the UK. Among them, the NT is currently the biggest organisation, owning 248 000 hectares of land, 600 miles of coastline, and 200 buildings and gardens. Its membership is presently over 3 million. Of the NT's land, over 80% is farmed and over 2 000 tenant farmers have worked on agricultural production since its creation. The NT was founded in 1895 by three philanthropists to protect countryside, coastlines and building threatened by development and industrialisation. Its major activities are conservation of countryside assets like historic houses, beautiful landscape and nature.

The basic mechanism of a conservation trust is to collect funds from its supporters and invest these funds in order to conserve the resources of the countryside. The NT's total income in 2002/2003 was GBP 303.6 million,[3] an increase from GBP 182.4 million in 1998/1999. One of the main reasons for this increase was growing membership and increased income from its businesses. The share coming from tenant rents decreased however. Of GBP 303.6 million, rent from tenants was GBP 23.6 million[4] in 2002/2003, about 7.7% as opposed to 11.0% in 1998/1999. This decrease was largely due to decreases in tenant's income from agricultural production.

The NT holds the view that farming should not only produce food, but should also shape the landscape and provide environmental benefits. The continuation of farming is therefore an essential condition to maintaining the countryside. Unlike other landowners, the NT seeks to balance the financial viability of the agricultural business and the maintenance of environmental quality. In practice, however, recent low incomes from farming and the resulting reduced capacity to invest in the improvement of the farm environment have made for a difficult situation.

Non-commodity outputs addressed and the mechanism used

NCOs in question for the NT case include landscape, biological resources and habitats, and/or historical features, all of which are more or less linked to agricultural production practices. Consumers, *i.e.* the NT members, appreciate the environmental goods but not the commodity outputs. The NT aims to maximise environmental benefits from agriculture and minimise its negative effects.

In the UK, tenants are granted a relatively high degree of autonomy in their management under the relevant legislation that prescribes the rights of both landlords and tenants. The NT therefore does not impose any substantive conservation obligations on its tenants, although it attaches great importance to the conservation of its property. The NT, instead, considers that communication and education of its tenant farmers have a vital role in reconciling two objectives, *i.e.* profitable agriculture and conservation of the environment. Under this policy, the establishment of the Whole Farm Plan by individual farmers seeking both more income and farm environmental conservation is encouraged and assisted by the NT. The implementation of the plan is monitored by local NT staff on a yearly basis.

There are two modes of conservation practice: with payment and without payment. Local conditions among estates are so diversified that whether a practice is paid or not is determined on a case-by-case basis. Nonetheless, the governmental guideline of Good Farming Practice[5] is a tacit standard dividing the two modes. No extensive survey has been conducted, but it seems that about 70% of tenant farmers observe almost all elements of Good Farming Practice. These tenants have more or less undertaken positive conservation works with payment. Others, mainly due to low productivity linked to the characteristics of the land they are working have found it difficult to develop farming practices following the conservation guidelines for commercial farming enterprises.

The conservation practices with payment include, for example, specifying stocking, grazing, and cutting conditions, and maintaining hedges and other landscape features, which explicitly reduce farm productivity or require additional labour. These practices agreed in advance between the two parties are sometimes included in a tenancy agreement, but recently it has been recommended that these be prepared in a separate agreement under the current tenancy legislation. The former case needs an arrangement for reduced rent; a merit of the latter is greater flexibility in implementation. For example, where a part of the conservation work is suspended for unforeseen reasons, it is easier to amend the agreement because no one is bound by a tenancy agreement.

Conservation practices that receive no payment include manure application to farmlands or its storage, soil management to prevent erosion, and limits on pesticide use. These lead to the mitigation of negative effects of agricultural practices, which are also essential elements to improving the value of the countryside. These practices are not enforced by law, but are voluntarily performed.

Considering the recent financial difficulties in the agricultural sector of the UK, the NT is prudent in acquiring more agricultural lands. Farmland with normal productivity may become a financial burden. Each year the NT is solicited to purchase large areas of farmland, but it has actually purchased only some of this land area, prioritising its acquisition to donated lands or to land that has some ecological importance.

Institutional setting focusing on governmental policies

The Agricultural Tenancies Act 1995, which replaced the Agricultural Holdings Act 1986, currently governs tenancy agreements in the UK. A newly added provision of the Act stipulates that when a tenant improves the land, thus raising the value of the holding, this should be compensated by the landlord when the tenant leaves. The compensation payment would require prior consent by the landlord and must take into account prior financial contributions from the landlord or any government grant schemes. This new provision encourages tenants to undertake environmental improvement works to enhance the environmental value of their land. However, the vast majority of the NT's tenants are under the old tenancy regime. Nevertheless, in view of the NT's high environmental motivation, it seems clear that it would have undertaken measures to encourage environmental work and pay for it regardless of this development in tenancy legislation. Thus, the new legislation does not seem to have had an immediate impact on the NT.

The NT urges its tenant farmers to apply for external financial resources to implement conservation engagements. This includes government agri-environment programs such as the Country Stewardship Scheme. The NT's income from government grants was GBP 1.4 million in 1998, which was payable to tenants through the NT in the form of adjusted rents or conservation payments. However, funding seems to be insufficient to meet targets. In this case, the NT supplements the amount to accomplish the target set. This implies that the NT environmental target is likely to be higher than the government's.

There are special legal provisions under the National Trust Act. The National Trust Act 1907 provides the basis of trust activities, such as its objective and the basic rules for NT property. According to Dwyer and Hodge (1996), a salient feature of the law is that trust properties are inalienable except when expropriated by the government with the consent of Parliament, a rare occurrence. This arrangement gives donators a guarantee that donated properties will be permanently preserved. Secondly, NT property is mandated to be devoted to public benefit in addition to NT members, with the implication that free riders are not an issue. Nevertheless, the NT is not completely open to the public. For example, the NT gives some exclusive privileges to its members, such as entrance to its properties with lower fees. Therefore, in economic terms, the NT has two facets: it is a club that has exclusive benefits for its members and it is a public organisation which serves all UK citizens. This is confirmed by the fact that important NT decisions are made by its council, half of whose members are elected from external organisations and the remaining half from its own members.

Overall, charitable organisations are governed by a series of charity acts that regulate and protect their activities, including fund-raising, registration, etc. The NT must comply with these laws.

The National Parks and Access to the Countryside Act 1949, which has since been replaced by the Countryside and Right of Way Act 2000, was the first law that permitted public access to rural sites of beauty located on privately-owned farmlands and to other countryside resources. The current law improves access to farmlands by expanding the number of such areas while recognising that the conservation of nature and wildlife are important, as well as reinforcing the right of landowners to restrict access to their land.[6] Such laws have long been called for by UK society as a whole, and therefore find considerable support.

There also exist tax relief schemes for charities under the UK tax regime which have a positive effect on the promotion of trust arrangements. An individual who donates money to a charity can reclaim the sum at the basic rate tax. If individuals give regularly to a charity through direct deductions from their salary, their employer deducts this sum before calculating the total salary subject to income tax. In addition, there is no inheritance tax on bequests to charities.

Organisational arrangements

One advantage of trust-type consumption of NCOs is collective purchase. High transaction costs have been avoided by contracting between the NT's tenant farmers and the NT's members, which number over 3 million. To properly reflect member demand for the provision of NCOs, the NT should provide sufficient information on goods, lands and conservation works and should know its member' preferences. The NT's council and executive council are forums where NT members can convey their preferences, although the forums do not discuss details.

Measures to promote NCO transactions

Recent rapid growth in the number of NT supporters[7] and the resultant big increase in its income can be attributed to several factors according to NT staff. First, there has been increased interest in UK history, triggered by several television series highlighting the history of various families and the legacies they left. These programs seemed to call to the attention of viewers that NT plays an important role in conserving historical homes. Another factor is growing environmental awareness in the UK, and the world in general. Such a factor is beyond the control of the NT, but the introduction of information technology, by which the NT provides information on its activities and recruits members on-line facilitates both the NT's and potential members, although the contribution of such technologies is not measurable.

Costs and difficulties encountered in getting the mechanism to work

As mentioned earlier, the financial viability of agriculture and the maintenance of the NT's agricultural lands are at risk due to recent low profitability in the sector. Dwyer and Hodge (1996) argued that the continued decline of farm income would lead to negative rent in the future in order to maintain local farming systems, although the rent from farming was the major source of income for the NT in the past. The NT, as owner of lands as well as being a conservation charity organisation, must assist in sustaining its tenant's financial viability and simultaneously enhance environmental quality. There may be trade-off between these two objectives. Hearn (1992) estimated that the average reduction of farm productivity due to conservation work was equivalent to GBP 5.9 per hectare, reflected in rent decrease. The NT, to overcome this situation, has urged its farmers to implement the Whole Farm Plan, which mainly aims to raise farmer's income by diversifying their income sources and promoting conservation work under the prospective improvement of their financial situations.

Of the NT's agricultural lands, about 60% is upland where livestock is the major income source, and where productivity is inherently lower than that of flat areas. Meanwhile, uplands generally form more attractive landscapes than do lowlands. Nonetheless, many upland tenant farmers are aged and it is very difficult to find successors who have the necessary skills and willingness to work under hard economic and physical conditions. As a result, the NT is likely to transform such lands to natural woodlands of biological interest. This would mark a change in the NT's future strategy on the management of its lands.

Assessment of institutional and other arrangements and conclusions

The NT collects funds from its members and allocates them for the conservation of countryside resources. The NT council, which represents members, is authorised to decide the allocation of funds among its properties, which include not only farmlands but also historical buildings and gardens which are major users of the fund. This mechanism works as a collective valuation of NCOs. On the production side, farm conservation payments are made to individual farmers who have developed a Whole Farm Plan which includes NCO conservation. This form of payment reflects the value of each NCO, thus avoiding ineffective payments being made to farmers who produce poor NCOs for example, but associated transaction costs are probably high. These valuation and payment mechanisms change as areas requesting conservation and the number of members change.

It is possible that the income level of NT members is higher than that of UK average.[8] Equity across income levels may thus not always be secured. In terms of geographical distribution, as the NT is accountable only to its members and there is no obligation to maintain fairness among regions, this is not an issue despite the fact that the NT is likely to be biased to conserve specific areas.

The NT's farmlands are conserved due to the law. In this sense, the mechanism is stable. However, recent difficulties governing the NT in the UK agriculture due to low commodity prices threaten the basic strategy of developing sustainable farming. This strategy could be particularly vulnerable in upland areas where agricultural profitability is lower regardless of higher environmental value. This problem is and will be crucial to sustaining trust-type NCO conservation. In this regard, agricultural policies significantly affect the financial viability and the environmental quality.

Notes

1. There is a sister organisation, the National Trust for Scotland founded in 1931. It has similar purposes and activities as the National Trust, but was established under separate legislation.

2. Aside from big trusts, such as the National Trust, the National trust for Scotland, Wildlife Trusts, and Royal Society for Nature Conservation, Dwyer and Hodge (1996) identified 122 independent "conservation, amenity, and recreation" trusts.

3. GBP 1.0 is equivalent to about EUR 1.49 as of March 2004.

4. This included income from agricultural grants and agri-environmental schemes.

5. The UK's regulations prescribe that farmers entering into agri-environmental commitments and/or in receipt of compensatory allowances in Less Favoured Areas Program respect 'Good Farming Practice'. Agri-environment payments are intended to compensate or provide on incentive for farmers to undertake measures that go beyond Good Farming Practice (Department of Environment, Food and Rural Affairs in the UK website).

6. Some 50 000 hectares of access is secured although around 500 000 in England and Wales are not permitted to be accessed, and further 60 000 hectares are permitted on an informal or *de facto* basis (Explanatory note to Countryside and Rights of Way Act 2000, 2000).

7. Membership was 2 million around 1990 and exceeded 3 million in 2002.

8. The membership fee per an adult per year is about GBP 30, but it changes by measures of payment and the number of joint membership with family, for example.

4. Easement arrangements in the United States (Type P3)

Abstract and classification

Agricultural conservation easement, as employed in the United States, is an agreement between a landowner and an NGO or a government authority concerning the permanent release of development rights with a view to maintaining the existing land surface to the benefit of the public. The landowner receives a lump sum payment for the sale of easement and/or a tax deduction as compensation for NCO production. This case study, which focuses on private initiatives, involves transactions between farmers and NGOs where no food commodities are involved. This case can be categorised as Type P3.

Background

Due to ever expanding urbanisation, much farmland has been converted to industrial or residential use. The result is a lack of open space, wildlife habitats, good landscape, etc., as well as reduced agricultural production capability. The conversion seems to be motivated by profitability. There are nevertheless farmers and members of the public who want to leave their farmlands for the next generation and who therefore seek to protect agriculture and its external benefits.

The idea of easement arrangement dates from the late 19th century, but it became significant with the enactment of the Uniform Conservation Easement Act in 1981. This Act enabled the imposition of stipulations at the time of the transaction that would put in place durable restrictions on the use of land to protect natural and historic resources. In comparison with land acquisition as a measure of land conservation, the use of easement has been rapidly growing because easements are less costly for NGOs and landowners still hold the title to the land on which the easement has been granted (Gustanski, 2000). About 1.4 million acres have been protected in the US by land trusts easement holdings, which include non-farmlands such as forests and wetlands (Land Trust Alliances, 1998).[1]

Non-commodity outputs addressed and the mechanism used

The ownership of land involves a number of rights and responsibilities attached to that land. The rights, for example, include the extraction of water, timber, and minerals, as well as the possibility to use that land for commercial and industrial purposes. Through easement arrangements, although the landowner transfers a part of his rights, he continues to own that land. The purpose of conservation easement is to have a legally binding agreement between a landowner and a qualified conservation organisation or public agency and in which the landowner voluntarily and permanently restricts development and future use of his/her land to ensure the protection of its conservation values, while still pursuing economic gains from the land. The landowner receives income and/estate tax reductions in return for selling or donating easement to government bodies or qualified private charitable organisations.

Agricultural conservation easement is designed to keep lands available for farming while limiting the construction of subdivisions and other non-farm development, and thus conserve the environmental output produced on farmlands. The Internal Revenue Code Section 170, which grants income tax relief to landowners donating easement to a charity, defines the conservation purpose as "… (ii) the protection of a relatively natural habitat of fish, wildlife, or plants, or similar ecosystem, … (iii) the preservation of open space (including farmland…) where such preservation is for the scenic enjoyment of the general public…". This indicates that goods conserved in easement arrangements should have the characteristics of public goods.

Non-economic factors seem to motivate landowners to apply to the conservation easement programme. Daniels (2000) states that, in general, landowners do so because they enjoy their land and the lifestyle it offers. They wish to pass them on to future generations who despite the relatively low rate of return from agriculture, and the value of farmlands as lots for residential and commercial uses, may also wish nevertheless to conserve the land.

The complex process of preparing and enforcing easement requires the expertise of attorneys, tax professionals, and environmental experts. Easement documents are tailor-made because land characteristics and circumstances are varied; the scope and goals for conservation easement are, accordingly, determined through case-by-case negotiation. As a first step, the goals envisaged by the two parties (easement suppliers and holders) must be reconciled through discussions assisted by a number of services which may include conservation NGOs and governmental agencies. Once agreement to embark on the easement process is reached, a baseline survey of the condition of targeted parcels of land is prepared in order to fulfil the federal taxation authority's documentation requirement. In particular, specific provisions to accomplish the desired conservation objectives must be drafted into the easement agreement (Diehl and Barrett, 1988). Although these provisions may restrict some activities, they do allow the landowner to reserve other unanticipated interests which could harm the holder's interest (Wiebe *et al.* 1996).

The next step is to appraise the property. A qualified appraiser, independent from both easement suppliers and holders, furnishes both the current market value of the property and the property value after an easement is placed. The difference in the two valuations is basically the price of the easement when it is sold at a fair market price. The easement may also be traded in a "bargain" sale[2] or donated, with a tax deduction applicable to both cases. Prospective easement holders, of course, prefer the donation as it is less of a financial burden for the holders, especially for NGOs who may have a relatively weak financial basis. Landowners are responsible for the preparation of these documents which cost several thousand US dollars (Ducks Unlimited, 2004).

Once these two steps are accomplished, the parties concerned select an easement conveyance strategy and arrange for compensation to the landowners. The next step is to record the easement in the office of the local recorder of deeds, during which the landowners must obtain title information including mortgage, if any, from the title examiner or attorney and disclose it. As long as the easement is attached to the land, all subsequent owners, whether they have purchased the land or inherited it, are obliged to abide by the easement provisions. If the landowners want to terminate the easement, the request has to be brought to court and is subject to its decision.

Monitoring generally carried out by easement holders is an important process to check the landowner's compliance with the agreed terms of conditions. American Farmlands Trust, for example, sends local staff every two years to all lands on which it holds easements. During the visit, which takes about half a day, the staff visit the site and fill out a questionnaire. In cases of violation of the terms of the easement, the easement holder obliges the landowner to correct the problems or face legal action.

Landowners may enjoy tax deduction benefits: income, property and estate taxes. The requirements for income tax deduction are: (i) an easement is donated (or sold at a bargain price) to a qualified charitable organisation, and (ii) the amount of donated easement is worth more than USD 5 000. The landowners deduct the value of the easement that is not actually paid by easement holders from their net income. The relevant federal tax law stipulates that up to 30% of adjusted gross income can be deducted and for six years at the most. This income tax deduction promotes the donation (or sale at a bargain price) of easement. The loss in income tax revenue is about USD 5 million annually (Small, 2000). Property and estate taxes (the former is payable by the present and future property owners, and the latter is payable by an heir) are determined based on the current fair market value of the property in question. Thus, once easement is in place, a reduction of the market value caused by easement land use restrictions leads to a corresponding reduction in taxes.

The case of Ducks Unlimited

Ducks Unlimited (DU), founded in 1937, is a non-profit private organisation and its mission is to conserve, restore, and manage wetlands and associated habitats for North America's waterfowl. Indeed, the primary cause of the decline in the population of waterfowl is loss and degradation of wetlands and adjacent uplands for nesting. DU has more than 1 million supporters across the United States, Canada, and Mexico and conserves more than 9.4 million acres of waterfowl habitats through various conservation measures including conservation easement.

DU, believes that conservation easement is the most effective land protection tool and therefore encourages landowners to sell or donate their land.[3] As of 2004, DU held 200 000 acres, including farmland easement in targeted areas identified as the most important for the protection of migratory waterfowls. DU's obligations as an easement holder are basically monitoring and enforcement of the terms of easement. A regional biologist visits a property on behalf of DU every year to check compliance. In addition to regular monitoring, DU assists landowners or land managers in maintaining environmentally healthy conditions by providing technical advice on request.

Institutional setting focusing on governmental policies

The federal and state governments' primary role is to make the rules for easement arrangements, which include the Uniform Conservation Easement Act (UCEA) approved in 1981, equivalent state laws, and several relevant tax codes. Following steps taken by several states that had already enacted conservation easement laws in the 1970's, the federal government enacted the UCEA to serve as a model for state legislation.[4] Its notable feature in the context of non-governmental approaches is that charitable organisations can become easement holders. This is promoted by tax provisions that allow tax deduction for easement donations.

In addition to developing rules, the federal, state and local governments have programs to purchase agricultural conservation easement. The federal government's Farm and Ranch Lands Protection Program (FRPP), created in 1996, provides matching funds to government agencies and charitable organisations. An important qualification to apply to the FRPP is that the applicant must have a conservation plan for highly erodible land, which is intended to prevent the loss of productive soil suitable for agricultural production. This program had protected more than 170 000 acres by the end of 2002. States and local governments also have equivalent programs. A survey in 2003 identified 46 state or local government programs to purchase agricultural conservation easement in 15 states, for an estimated total cost of USD 2 billion covering 1.8 million acres of farmland.[5]

Conservation NGOs can qualify for FRPP's and for other local government program financing to purchase easement. As mentioned above, the expenses for easement holders include costs for purchase, monitoring, and administration. The participation of NGOs in the FRPP enables these costs to be shared between the government and the NGOs, bringing mutual benefits.

Costs and difficulties encountered in getting the mechanism to work

A serious disadvantage for easements in general is the high transaction costs involving case-by-case negotiation, appraisal, monitoring and enforcement due to characteristics specific to individual players and lands to be conserved (Wiebe *et al.* 1996). To ease this complex procedure and reduce the costs, a standardised contract format might be preferable but there is a trade-off in that this would eliminate the exact description of duties and commitments and might therefore lead to a larger number of disputes. Governments have issued some guidelines to try to make the procedure more uniform and to reduce costs, but with limited impact.

There are two potential problems posed by the perpetuity aspect of easement: the landowner's incentive for permanent conservation may change over time as might the viability of agricultural activity reflecting changes in technologies and society. It is very costly to change the terms of an easement arrangement. However, only the landowner who first agrees to the easement can gain payment or income tax benefits. Successive landowners have no such incentive. Further, a change in the land surface within the agreed terms may not be accepted by an easement holder. This problem could result from insufficiently detailed individual provisions at the initial stage or from a difference in the interpretation of specific provisions. It would therefore increase transaction costs for dispute and monitoring.

In relation to the second point, McKee (2000) has shown that agricultural viability may be jeopardised if the landowner is not allowed to adapt to changes occurring in markets, technologies, and consumer's tastes and demands. In this event, the farm landscape or habitat of interest will not survive. The type of farmland conservation prescribed and agreed in easement documents is generally based on conditions prevailing at the time of signature, or on conditions in the foreseeable future, but long-term, unanticipated changes may be difficult.

Assessment of institutional and other arrangements and conclusions

Abandoning the right to convert to non-agricultural use of land is motivated by a desire for the continuation of agriculture and for associated NCO production. The payment from NGOs, if any, combined with tax relief granted from government for easement acquisition can be seen as compensation for the production of public benefits. The payment is based on the market price of land which is different from other mechanisms such as agri-tourism or market price premiums that value the NCO on the basis of the consumer's evaluation. A lump sum payment is made for easements while payments are made for other measures through the price and depend on the existence of the relevant markets.

Efficiency requires the valuation of NCOs and corresponding payment to farmers. In the case of easements, valuation and payment are realised in a unique way compared to other NGAs. A message from easement holders (consumers of NCOs) to suppliers (farmer landowners) translates into the continuation of agriculture, which maintains the landscape and ecology. This does not necessarily lead to an improvement in the quality of the NCO. The price of the NCO is that of the easement which is determined as the difference before and after the placement of the easement. This pricing process is transparent. Payment to the farmer-landowner is made by all easement holders or partly by government through some form of tax relief. In the latter case, it is essentially the government that is the purchaser and we are no longer dealing with a NGA.

If the NGOs in question are not rich men's clubs, and are open to all regardless of income level, a high degree of equity can be achieved. However, a free-rider problem is still inevitable as non-NGO members can enjoy the NCOs. Further, the NGO arrangement is effective in conserving specific NCOs that are favoured by the group of people comprising the NGO based on their own preferences which may not correspond to wider preferences in society and which may mean that the conservation activity is concentrated in particular areas or regions. Easement arrangements are usually employed and effective for a piece of land, which is not so large, and is owned by a single landlord. Moreover, the land has potential for development and therefore a need for conservation, because it is near an urban area. These conditions alone limit the area for which this type of arrangement is relevant. Therefore, an equity question across regions is unlikely to arise.

Perpetuity is secured in easement documents, which should generate stability of NCO provision. Mechanisms guaranteed by legislation may have a high degree of institutional stability. However, many easement arrangements do not have a long history so that potential problems in this domain are difficult to anticipate at this time.

Notes

1. According to the Land Trust Alliance (1998), US Land Trusts, composed of 1 218 local land trusts, conserve 3.1 million acres in total, among which about 43% is protected by easement, 26% by owners, and the remaining 31% by third parties.

2   When a landowner sells the easement to a registered charity at a bargain price, he will enjoy both reduced tax benefits and payment through the sale.

3. The vast majority of DU's conservation easement is donation.

4. Mayo (2000) indicates that among 50 states in the United States, 21 states adopted the Uniform Conservation Easement Act with or without modifications, and therefore there is high degree of uniformity and consistency in their easement legislation. Another 25 states have specific legislation.

5. The survey, the National Assessment of Agricultural Easement Program, is a joint project of American Farmland Trust and the Agricultural Issue Center of the University of California, Davis.

5. Agritourism and landscape conservation program in Austria (Type P3)[1]

Abstract and classification

The "Landscape Conservation Program" is an arrangement where beneficiaries (tourists and the community) of rural landscape pay a fee to registered farmers who adhere to specific landscape cultivation guidelines so as to conserve farm landscapes. Because this transaction takes place between a private organisation administering the program and the farmers, this case can be categorised as a club provision, and thus as Type P3, where an organisation collectively conducts transactions with farmers and no agricultural commodities are involved. However, a notable feature of this arrangement is that individual tourists and the tourist industry participate in this program, which can be categorised in Type P1.

Introduction and background

The community of Weißensee is located in the alpine region of the province of Carinthia in the south of Austria. Its elevation is more than 900 meters and the area is topographically characterised by steep slopes, many of which are covered by forests. In the centre of the community lies an alpine lake, the Weißensee, which is 11.5 kilometres long and up to 900 meters broad. The appearance of the countryside is stamped by agriculture, as it is for many Austrian mountain areas. The area of Weißensee is currently managed by 26 mountain farmers, most of whom have switched to organic farming in recent years.

The economic structure of this area is unbalanced, with tourism dominating the community for decades. With more than 450 000 overnight stays per year and only 800 inhabitants, Weißensee is one of Austria's most tourist-oriented communities. One hundred hotel keepers and 102 private lodges provide 4 000 beds on aggregate, and two campgrounds provide additional space for several hundred people. Approximately 80% of overnight stays occur during the summer period. Agriculture is closely connected to the tourist industry. First, farmers often lease apartments or guest rooms to vacationers; secondly, tourists demand regional agricultural products; and thirdly, the (agri-) cultural landscape represents an important input factor for the production of tourist services.

As in all Austrian mountain areas, however, the economic role of agriculture has been dramatically reduced over of the last several decades. Due to restricted mechanisation connected with topographical disadvantages, productivity growth of mountain farms has been significantly lower as compared to competitors in lowland areas. As a consequence, decreasing commodity prices have reduced agricultural income of mountain farmers in the Weißensee area and elsewhere, a development which has not been attenuated even with the direct financial compensation for less-favoured areas provided at the European and the national levels. Many of the "marginal producers" in these regions have closed their business and have migrated to other areas.[2] Others have started operating a second business (leasing guest rooms for example) and managing their farms part-time. Today, 26 farms continue to operate on a part time basis. The decrease in agricultural employment not only has had direct economic consequences on the region, but it has also threatened the continued provision of agricultural landscape-cultivating services.

Green tourism

Several years ago, the community started implementing the concept of green tourism with a focus on preserving nature and promoting environmental awareness among vacationers and local residents. Measures that have been undertaken include the ban of constructing new buildings in valuable and sensitive zones pertaining to natural scenery, the cultivation of the rural landscape, no further increases in the number of tourist beds, the reduction of local traffic through accompanying steering mechanisms, and the promotion of architecture that is typical of the region. The overall idea was to qualitatively improve the tourist infrastructure with a clear focus on the local environment.

With involvement of the local population, the Weißensee administration has developed guidelines for the future development of the region. The mission statement generally acknowledges the interdependence between agriculture and tourism, and the continuation of current agricultural use of the rural landscape as the basis for production of agricultural commodities. Several objectives are listed in the subsection "agriculture and the countryside". These include:

- increasing the value and the importance of agriculture by providing both material support and ideas for further development.

- promotion of direct commercialisation of agricultural products (organic commodities).

- creating awareness of the singularity of the countryside.

- moratorium on significant changes to landscape; and

- no further development of mountain pastures through forest roads and trails.

These items clearly illustrate the focus on agriculture's responsibility for the cultivation and enhancement of the rural landscape and the importance of multifunctional agricultural outputs.

Description of the non-commodity output (NCO)

The most important NCOs in Weißensee (as in numerous other Austrian regions) are primarily centred around the preservation of a typical agricultural landscape for the enjoyment of tourists and residents. The scenic landscape with its environmental amenities is indispensable for the prosperity of tourism in this region. Even though the potential for a rural area to provide countryside benefits depends on ecological and geographical factors (the presence of species and habitats, the area's capacity to regenerate and generate new habitat, climatic and geomorphological conditions, etc.) many landscape benefits arise as (joint) products of agricultural production. The relevant landscape-cultivating services in the Weißensee region include mowing, maintaining the rural trail and road network, preserving the vegetation along the waterside, and cultivating alpine pastures, all of which could be provided independent from agriculture. Moreover, the diversified arrangement of groups of trees, hedgerows and brushwood contributes to the preservation of species. These services exert a positive influence on the utility of those who spend (leisure) time in the region. They perfectly meet the multifunctionality properties of agriculture: jointness of production, externalities and market failure, and public good characteristics.

The institutional setting and organisational arrangements

The objective of the so-called "Landscape Conservation Program" in the Weißensee area — a by-product of the green tourism concept — is to protect the rural landscape and ecology and to prevent farmers from quitting their business. For this purpose, a private club has been founded, the so-called "Landscape Conservation Organisation". Farmers who wanted to participate in this program had to register with this private association. The Organisation has set up comprehensive production and landscape guidelines to be followed by farmers seeking monetary NCO compensation. Based on individual farm characteristics (see below), farmers earn "scores", with the total number determining the compensation received from the Organisation for their NCOs. All 26 farmers in Weißensee participate in this program.

The point system and the pay-out of funds

The original idea was to distinguish single areas of 300 hectares of grasslands under cultivation according to their "societal values". In the first draft of a point rationing scheme, grasslands mowed manually were given priority, followed by areas that allow the use of engine mowers or tractors. The overall score allocated to a single farmer therefore depended on the relative percentage of areas falling into these different grassland categories. In 2001 this scoring system was adapted when a new classification which included mountain farms was introduced in Austria. The so-called "Berghöfekataster (BHK)" covers the objective degree of difficulty in cultivation at the farm level und translates these managing difficulties in mountain areas into a point system.[3] Sixteen single criteria, such as the inclination of cultivated areas, elevation, classification numbers for climate and land, accessibility of the farmyard, distance of the farmyard to local infrastructure (bus stop, railway station, municipal offices, etc.) and individual farm characteristics were merged into a single score.[4] The highest possible score is 570 points with more than 50% reserved for the steepness of the inclinations and the number of very small parcels of land separated from one another. The BHK scores of mountain farms in the Weißensee region lie between 105 points and 207 points (medium to high degree of difficulty in cultivation). The payment to an individual farmer depends on the multiplication of his BHK score with the number of hectares under cultivation. However, for entitlement of payments, a contract which imposes a number of conditions needs to be signed between the individual farmer and the Landscape Conservation Organisation. The most important of these conditions are:

- grassland must be mowed at least once a year.

- cut grass must be removed from all areas.

- the use of chemical fertiliser, pesticide, herbicide, and fungicide is not allowed.

- afforestation of areas under contract is prohibited.

- livestock density must lie between 0.5 and 2.0 livestock units; and

- compliance with specific regulations for certain types of grassland (marsh areas).

Farmers who keep these conditions receive payments, however scores will be deducted in case of non-compliance with single items. The managing board of the organisation is in charge of monitoring. All areas under contract are controlled every year within on-site inspections. In addition to the payments for mowing the grassland, farms receive monetary compensation for managing marsh areas, growing potatoes, and for husbandry all of which contribute to the diversity of the countryside. These activities receive EUR 290.70 per hectare in support payments. Table II.1 shows the local Landscape Conservation Compensation Scheme for the farmers in the Weißensee region for the year 2001.

The overall monetary compensation for the year 2001 was EUR 43 600. A farmer therefore received EUR 1 677 on average. Given a total of 294 ha of grassland and taking into account the payments for marshland, husbandry and potatoes, the program has paid EUR 0.98 per hectare and per BHK point. The total amount of monetary funds is not fixed in nominal values; it can be adjusted according to cost-of-living indices.

Table II.1. Landscape Conservation Compensation Scheme

Farm Number	BHK points	Area [ha]	Marsh areas, husbandry, potatoes [ha]	EUR	Total [EUR]
1	105	18.19	-	-	1875.49
2	106	6.97	0.60	174.41	899.90
3	107	9.33	0.88	255.81	1236.10
...	...	...	...	...	...
13	129	9.05	1.00	290.69	1 437.08
14	130	12.27	0.61	177.32	1 743.64
15	132	9.93	2.03	590.10	1 877.21
...	...	...	...	...	...
24	182	16.47	0.09	26.16	2 969.61
25	185	6.29	0.22	63.95	1 206.60
26	207	10.88	-	-	2 211.52
Total		294.00	14.00	4 075.50	43 600.00

The collection of funds

The Organisation's revenues are collected among tourists spending their vacation in the area. However, the tourists' payments do not represent direct private contributions to landscape cultivation. A local tourist fee of EUR 1.38 (peak season) or EUR 1.16 (off season) per person and per night is imposed on vacationers. A certain fraction of this community fee — 7.3 cents per overnight stay for persons 18 years and over — is directly transferred to the Landscape Conservation Organisation for compensating landscape cultivation and enhancement activities. In 2001, the total amount of these tourist fee revenues was approximately EUR 25 500. Moreover, the Organisation received additional transfers of EUR 18 100 from the community budget. A simple incidence analysis makes clear that a big fraction of overall payments is unambiguously created in the tourist industry since the sector is not reimbursed the whole amount of the imposed tourist fees. However, the administrative setting indicates that the local community is involved and the program, therefore, also includes quasi-market components (OECD, 2001b).

Cost and difficulties

Table II.1 shows that the financial cost of the program runs up to EUR 43 600 on aggregate. Transaction costs in the form of administrative costs are low since both the work of the Organisation's managing committee and the monitoring activities are provided by volunteers who do not receive any payment in return. Non-personal overhead (*e.g.* office equipment) is covered by the Weißensee municipal office. The only person paid is the cashier who receives a salary depending on the number of hours worked for the Organisation.

The availability of funds in the long term remains an open issue. The payments are granted voluntarily and restricted (public) budgets may jeopardise the program in the future even though the validity of the project has not been restricted to a certain time period.

Another problem concerning the relation between agriculture and tourism is that until now the agricultural sector has not sufficiently succeeded in promoting and positioning its organic farm products. Due to competition with traditional farm products and the high price elasticity of agricultural demand, farmers have not had the opportunity to internalise landscape services into their prices for organic commodities.

Finally, there is one more negative point concerning contact between agriculture and tourism: the turnout of liquid manure which causes the negative externality of unpleasant odours. From an agricultural perspective, the perfect point in time for this turnout is in mid-summer, which is the peak tourist season. However, the farmers have begun to use a mixture of herb essences to reduce the smell nuisance of liquid manure. This has alleviated the problem significantly.

Assessment and conclusion

The Weißensee Landscape Conservation Program was introduced 12 years ago. Since then, several beneficial consequences for the appearance of the landscape have been observed.

The landscape has been kept open, and grassland areas which were once threatened by overgrown trees and bushes, are currently managed. The agricultural cultivation of the waterline guarantees a perfect view of the lake. The increase in forest cover has come to an end since critical parcels of land, such as the edges of woods, have been mowed. Moreover, regular inspections of the lake have certified best water quality for a series of years. This achievement is at least partly influenced by forbidding the use of chemical fertilisers and pesticides.

There is a broad consensus in the local population about the necessity of the measures undertaken as the interdependence between tourism and agriculture is generally acknowledged. The program fulfils non-economic purposes as well since it contributes to social cohesion in the community. Irrespective of their political affiliation, the deputies in the municipal council voted unequivocally in favour of the landscape conservation program including compensation payments to farmers. The vacationers also express their appreciation of the chosen policy and the local tourist association supports the whole program. The concept of green tourism is a success for the region, and it is recognised that accompanying measures are needed to support extensive agricultural production. Farmers generally accept the contracts and fulfil their ecological requirements, even though they would prefer higher commodity prices.

It seems the program is economically efficient in so far as externalities are internalised (an increase of external benefits and a decrease of external costs) at relatively low cost.[5] This type of local compensation for agricultural NCOs may serve as a promising alternative, or at least as a supplement to EU and national policy measures in support of rural and/or less favoured areas. The model is appropriate for tourist intensive communities in particular as these municipalities dispose of significantly higher budgets as compared to communities in other regions. As an example, more than 50% of community revenues in Weißensee have been raised in the tourist sector. We argue that compensation payments being transferred to farmers are born by individual preferences for multifunctional agricultural outputs, and these aggregated preferences are expressed by the outcome of Coasian negotiations between suppliers and consumers (at the local community level and often including certain regulatory tools).

However, a transfer of this program to other, probably larger, areas might lead to significantly higher transaction costs if both administrative efforts and monitoring activities can no longer be provided exclusively by volunteers. Finally, a better accompanying marketing strategy seems recommendable. Tourists are not yet systematically informed about the specified use of part of the tourist fees.

This case study attempted to illustrate that the order of magnitude of local compensation payments represents a significant contribution to the farmers' income situation. The payments, however, are obviously not high enough to keep the suppliers of the rural landscape in business. More funds will be needed in the future for the sustainability of agricultural NCO provision.

Notes

1. This case study was prepared by Gerald J. Pruckner of the Department of Economics and Statistics, University of Innsbruck.

2. The number of mountain farms in Austria has decreased by more than 12% per decade since 1980.

3. The BHK is also the basis for compensation payments to farmers in less-favoured areas under the EU Rural Development Program and for certain premiums being paid under the Austrian Environmental Program for Agriculture, AEPA.

4. See Hovorka (2002) for details.

5. A reliable assessment of efficiency, however, would require a thorough analysis as to whether people's (marginal) willingness to pay for agricultural NCOs exceeds marginal cost of provision.

6. Tourist train in Norway (Type P3)[1]

Abstract and classification

In this case study, a group of farmers whose farms are located alongside railway tracks used by a tourist train is paid to conserve the landscape of the countryside. This case study can thus be categorised in Type P3, where individual tourists indirectly and through a private organisation pay to the farmers' conservation of the landscape and where, in addition, no food commodities are involved in the transaction.

Background

Every year about 400 000 passengers travel on the Flåm Railway (www.visitflam.no), one of the most visited tourist attractions in Norway. The 20 km railway line runs along the narrow Flåmsdalen valley in the inner part of the Sognefjord from the village of Flåm at sea level, up through the narrow Flåm Valley to Myrdal railway station 856 meters above sea level, where it is connected to the Bergen railway. It is the steepest line in northern Europe with standard gauge, and 80% of the line has a gradient of 55%. The Flåm railway was completed in 1940 and is built on narrow ledges in the mountainside with tunnel loops through a typical western Norwegian countryside of lush vegetation, waterfalls and steep mountainsides. Glaciers, green pastures and birch forests frame the Flåm River, which the railway crosses three times (Thune, 2002). Tourists come to experience the breathtaking views of this mountain and fjord landscape, but due to a decrease in the population, farms and grazing animals (especially goats), the birch forest has grown to such an extent that today it conceals many of the great views. At present, less than 400 persons live in Flåm and the Flåm valley, but there are still 19 farms with agricultural production. Most of these have sheep, in addition to other production, while only six farms have goats (three of which are run as a joint operation).

Goats and kids are better than sheep in clearing brush and bushes and can be used to prepare pasture for sheep. Sheep are financially more attractive to farmers due to the difference in the general governmental transfers and market prices for the products, and have replaced goats on many farms. This, in turn, has led to an uncontrolled growth of birch forests. The manager of the railway, Flåm Development Ltd., has realised that goats and kids can be effectively used to open up the views from the railway and has played an active role in a project that combines a non-governmental approach (NGA) and governmental support to farmers for providing this non-commodity output (NCO). This case study can be seen as an example where general government transfers do not fully compensate the farmers for producing external benefits in terms of agricultural landscape, and where a combination of transfers from the private beneficiaries of this public good and specialised governmental transfers provide farmers with the incentive to produce a more efficient amount of this NCO.

Description of the non-commodity output (NCO)

Grazing goats provide NCOs in terms of keeping the vegetation down so that the tourists can enjoy the views from the railway and the parallel "Rallarvegen", an old railway construction road now used by tourists for walking and biking. In addition to improved views, the tourists can also enjoy the experience of viewing and getting close to the goats, which can also be classified as an NCO.

In 2000, three goat farmers, in co-operation with the private landscape planning and management firm "Aurland Naturverksted", initiated a project to control grazing with goats to manage the cultural landscape in the Aurland community, which would also improve the tourists experience of the Flåm Railway. Flåm Development Ltd. quickly saw the positive effects for the Flåm Railway and offered to support the project on the condition that the grazing areas were selected so as to improve the views from the Flåm Railway. Aurland Naturverkstad and Flåm Development Ltd. surveyed the area from the railroad in 2000 and identified the most important areas for improving the views from the railway.

The private company Flåm Development Ltd., took over marketing and product responsibility for the Flåm Railway from NSB AS (the Norwegian State Railways) on 1 January 1998, making it one of Norway's few privately-owned railways. NSB is still responsible for security, handling and the actual running of the trains. The railway tracks are owned and managed by the Norwegian National Rail Administration (Jernbaneverket). Flåm Development Ltd is owned by the municipality of Aurland (which Flåm is a part of), through the parent company Aurland Ressursutvikling AS the Savings Bank of Aurland, SIVA (The Industrial Development Corporation of Norway) and OVDS (which is one of the two coastal express companies).

The project lasted three years (2000-2002) and involved three goat farmers with a total of 100 kids the first year; 74 goats and 93 kids the second year; and 59 goats and 101 kids for the third year. All farmers had goats before the project started.

The "new" landscape that resulted is a NCO that is both non-excludable and non-rival (at least to the point where the crowds of tourists start reducing the recreational experience each of them has from enjoying the landscape). While tourists travelling on the Flåm Railway pay indirectly (or directly if Flåm Development Ltd. implements a planned environmental tax of NOK 1 per trip), the bikers and hikers along the Rallarvegen would, literally speaking, be free-riders! Other tourists and the local population also benefit.

The total budget of the project in 2000 was NOK 88 288 (NOK 1 = 0.12 euro). NOK 75 000 (85%) was financed by Flåm Development Ltd., while Aurland Naturverkstad financed the rest from its own funds (Table II.2). While the activity in 2000 was based solely on the efforts of the three farmers, the increased budget for 2001 and 2002 allowed for one person to be employed fulltime as a landscape manager, responsible for clearing trees and bushes, putting up fences, and controlling these fences as well as the grazing goats and kids. In 2001, the budget increased to NOK 342 484, out of which NOK 135 484 (40%) came from Flåm Development Ltd. The rest consisted of funds from the Aurland community administration for cutting trees along the main road (which they manage) (NOK 32 000), and government transfers; NOK 75 000 in specific government transfers for preserving agricultural landscapes (STILK) and NOK 75 000 in district development funds (BU) from the Sogn and Fjordane county administrations. In 2002, the total budget was NOK 273 367. NOK 120 367 (44%) was financed by Flåm Development Ltd. The remaining NOK 150 000 consisted of government transfers which was evenly split between STILK and BU funds (Aurland Naturverkstad 2001, 2002, 2003).

The transactions costs in terms of consultancy fees for the project management done by Aurland Naturverkstad was 25% of the annual budget for both in 2001 and 2002 (which was down from about 40% in the first, start-up year, out of which ¼ was self-financed by Aurland Naturverkstad).

None of the farmers increased the number of goats as a result of this monetary incentive, but one farmer kept the goat kids over the summer, when they would otherwise have been slaughtered and sold in the spring (when they fetch the highest prices). The annual reports from the project (Aurland Naturverkstad 2001, 2002 and 2003) show that there were two main driving forces behind the establishment of the project: *i)* An offer by the firm Vestlandske Salslag to increase the price of kid meat, and *ii)* the economic support from Flåm Development Ltd. to undertake fencing, clearing and cutting of trees, and looking after the grazing goats and kids.

Table II.2. Project Budget (2000-2002)

Year	Total budget (NOK)	Details (NOK)	(%)	Sources of funds
2000	88 288	75 000	85	Flåm Development Ltd.
		13 288	15	Others
2001	342 484	135 484	40	Flåm Development Ltd.
		32 000	9	Aurland community
		75 000	22	Government transfer for preserving agricultural landscape (STILK)
		75 000	22	District development funds (BU)
2002	273 367	120 367	44	Flåm Development Ltd.
		75 000	27	Government transfer for preserving agricultural landscape (STILK)
		75 000	27	District development funds (BU)
		3 000	1	Others

Source: Aurland Naturverkstad 2001, 2002, 2003.

Institutional setting and the role of governmental intervention

In 2001 and 2002, each of the goat farmers received NOK 100 per goat and kid per year to compensate for the additional workload that the controlled grazing implied. This compensation was financed by the governmental district development funds (BU) from the county of Sogn and Fjordane. Aurland Naturverkstad (2002 and 2003) provides calculations showing that an additional landscape management subsidy of NOK 200 per goat and NOK 100 per kid is needed to make goat farming as profitable to farmers as is sheep farming.

Organisational arrangements

The private landscape management firm Aurland Naturverkstad was found to play a crucial role in the success of the project and the provision of NCOs. They were important both as co-ordinators between farmers themselves (as a neutral, scientific body), and between the farmers and the private and governmental funding sources. They also played an important role as a scientific advisor and catalyst to secure both government transfers and the success of this NGA. Aurland Naturverkstad´s motivation for initiating this project was originally a genuine, academic interest in using goats for landscape management, but it also seems likely that they saw this project as a way to show both local and national agricultural authorities how their theories could be implemented in practice, and that firms like theirs were crucial for a successful implementation.

Other measures facilitating transactions

Aurland Naturverkstad also contacted the Norwegian Rail Administration, which had cleared the vegetation close to the tracks for security reasons (*e.g.* to avoid fires). They agreed to finance the construction of fences towards the tracks in the areas designated for the grazing project, realising that this project would also benefit them in terms of reduced needs for manual cutting of the vegetation near the tracks that they perform every fifth year.

Costs and difficulties encountered in getting the mechanism to work

The tragic death of the most active of the three farmers involved initially had a negative impact on the project in 2001. However, a new farmer became involved in 2002, so the total number of farmers remained three during the project period.

Even if the effect on the landscape was very good, Flåm Development Ltd. ended their involvement in 2002 as they felt the transaction costs, in terms of the high costs of hiring people to clear vegetation and for fencing, was too high to make the project efficient. They instead signed a three-year contract (2003-2005) with the Aurland Rifle Association at a cost of NOK 100 000 annually (Flåm Development Ltd plan to increase the amount for 2005). In 2003, the rifle association primarily cut down the birch forest, and Flåm Development Ltd has now realised that although this non-governmental organisation (NGO) is well suited for clearing forests, it is not for the co-ordination of a grazing program due to their lack of scientific expertise on grazing and more general expertise on co-ordinating farmers, overseeing the grazing project, and securing specific governmental funding (STILK and BU funds). They are, therefore, currently considering hiring Aurland Naturverkstad again as a co-ordinator to continue the original project.

Assessment and conclusion

The results from this case study should be transferable to other countries with similar broad and general government transfer schemes, and where private firms depend on views of agricultural landscape for their profits. See, for example, Hackl and Pruckner (1997) for an economic analysis of a case where hotel owners are paying farmers (on top of governmental transfers) to maintain a certain type of agricultural landscape. However, the broad and general government transfer mechanism in Norway (*i.e.* the "area and cultural landscape transfer "mechanism) is very different from the transfer mechanisms of many other European countries, which have narrower and more specialised mechanisms aimed at maintenance and reconstruction of agricultural landscapes. Overall, goat and sheep farming in Norway would not be profitable for farmers if they did not receive these general transfers. Thus, the success of the private mechanism described here depends on the existence of the general government transfer mechanism.

A mean, annual willingness-to-pay (WTP) of about NOK 1 from each of the 417 540 passengers in 2003 on the Flåm Railway (Flåm Development 2004) is sufficient to recover the costs of this project. Flåm Development Ltd. now considers introducing a private tax of NOK 1 (or € 0.12) for each tourist taking a trip on the railway. Note that the general government transfers need to be added to come up with an estimate of the WTP needed to offset the total social costs of this project. There are no studies of the WTP among the tourists on the Flåm Railway, but Hackl and Pruckner (1997) report a mean, annual WTP of € 0.70for the preservation of agricultural landscape among tourists in Austria from a Contingent Valuation survey in 1991 (Pruckner 1995).

Overall, the project would contribute to increased equity as it converts external benefits to income for the farmers providing these additional positive externalities for tourists. Note that these private transfers are in addition to the general government transfers to farmers for producing agricultural landscape. However, there would still be free riders in terms of foreign hikers and bicyclists on the "Rallarvegen" road parallel to the railway road, and additional foreign tourists visiting Flåm and the Flåm Valley that do not travel on the Flåm Railway. The local population and tourists from Norway are not free riders as they pay the farmers indirectly through general taxation.

The stability of this project to provide NCOs seems to be crucially dependent on (in random order): *i)* economic incentives provided to the farmers to compensate them for their production of NCOs; *ii)* perception of the magnitude of the transaction costs among the contributing non-governmental bodies, *iii)* commitment by the private firms that provide the NCOs, *iv)* a neutral and scientific coordinating body, and *v)* the success of the project in terms of producing NCOs.

Note

1. This case study was prepared by Ståle Navrud, Department of Economics and Resource Management, Agricultural University of Norway. Project leader Morten Clemetsen at Aurland Naturverkstad and managing director Olav Lühr at Flåm Development Ltd. provided data and answered questions and constructive discussions.

7. Consumer movement: *Chisan-chishou* in Japan (Type P4)

Abstract and classification

Chisan-chishou is a growing and popular consumer movement in Japan which aims to promote the consumption of locally-produced foods. Their intention is to promote safe foods and conserve local agriculture, the rural environment and economy. The exchange of agricultural commodities resulting in the conservation of NCOs takes place between organisations or individual consumers on the one hand and farmers. Therefore, this case can be categorised as Type P2 (transactions with individual consumers), or as P4 (transactions with organisations). This case study highlights the latter.

Background

In the 1970's, groups of consumers concerned about the negative effects of agricultural chemicals on humans and the environment, as well as the growing percentage of imported foods began direct growing and purchase of agricultural products by contracting farmers. This activity was called *teikei*, a direct translation, meaning partnership. The general principles included developing a production plan in accordance with the desires of consumers, who should purchase all outputs produced under the plan, and that prices would be mutually beneficial. This concept was adopted by other countries, including the United States where it became known as "community supported agriculture (CSA)". In the 1980's, however, this activity began to decline as organic products markets grew with supermarkets as well as individual retailers.

> **Box II.1. What is community supported agriculture (CSA)?**
> (*Cited from* **University** *of Massachusetts, 2000*)
>
> CSA is an innovative and resourceful strategy to connect local farmers with local consumers, to develop a regional food supply and a strong local economy, to maintain a sense of community, to encourage land stewardship, and to honour the knowledge and experience of growers and producers working on small to medium farms. CSA is a unique model of local agriculture which began 30 years ago in Japan. A group of women concerned about increasing food imports and the corresponding decrease in the farming population initiated a direct growing and purchasing relationship between their group and local farms...The concept travelled to Europe and to the U.S., where it was given the name "Community Supported Agriculture" at Indian Line Farm, Massachusetts in 1985. As of January 1999, there are over 1 000 CSA farms across the U.S. and Canada.
>
> CSA is a partnership of mutual commitment between a farm and a community of supporters, which provides a direct link between the production and consumption of food. Supporters cover a farm's yearly operating budget by purchasing a share of the season's harvest. CSA members make a commitment to support the farm throughout the season, and assume the costs, risks and bounty of growing food along with the farmer or grower.

From the 1960's, the urban population expanded and much farmland surrounding urban areas changed to residential or industrial use. These farmlands had been the main suppliers of vegetables to urban residents. Consequently, vegetable supplies became unstable and there were wide fluctuations in prices. To resolve this problem, the Japanese government developed policies to promote mass production of vegetables across rural areas and mass market-oriented strategies to supply big cities. As a result, many vegetable growers developed monoculture production and distribution systems because they were less costly to produce and distribute, in addition to generating greater profits. Meanwhile, it was reported with some surprise that many vegetables in rural markets came from cities' wholesale markets. Originally, local markets used to have plenty of fresh products harvested from nearby farms.

In the 1990's, in addition to the original concern about food safety, consumers, especially those living in rural areas, also became concerned about the decline in local agriculture and the irrational distribution systems. These concerns motivated the birth of a new movement called *chisan-chishou*, which means the promotion of local production by local consumption. In the course of the development of *chisan-chishou*, various concepts have been added. For example, consumers who are interested in global warming have introduced the concept of food mileage[1] to reduce fuel consumption for food transportation. Meanwhile, the trend towards the rediscovery of local foods and cuisine that had almost disappeared in modern living has merged with *chisan-chishou*.

Non-commodity outputs addressed and the mechanism used

What consumers seek from *chisan-chishou* are largely safe foods, the revitalisation of the local economy, and conservation of the local environment. Safe food cannot be considered as an external benefit of food production. Nonetheless, the production process, *e.g.* low application of chemicals, could be of benefit to the farm ecology, which is a part of the local environment. For the local economy, the expansion of local agricultural production and the corresponding spill-over effects seem to be of major interest to those who benefit from them. While this cannot be defined as a positive externality because the effect arises through market transactions and no market failure is involved, local production may produce benefits through farm employment, such as an even distribution of population, or reduced per capita costs of public service in rural areas (OECD, 2001a), although to what extent these benefits are important is unknown. Lastly, the rural environment has deteriorated mainly because of the use of chemicals and the increasing areas that are abandoned or set aside. The improvement of landscape is an NCO in this case.

All farmers, including non-*chisan-chishou,* contribute more or less to the local environment and economy. But, by diversifying the crops grown or by pursuing environmentally favourable practices more than other farmers, the *chisan-chishou* farmers may add to environmental quality. They incur higher production costs leading to higher consumer prices, although to date there seems to have been no extensive price investigations.

Chisan-chishou is the generic term for a social movement having various activities and groups across Japan such as CSA, U-pick up, a road-side stand, ordinary retailers dealing with local products, and the provision of school lunches made from local products. *Chisan-chishou* movements are initiated by private voluntary groups or individuals with or without local government assistance. Their objectives are diverse but most of them can be integrated into the conservation of local agriculture and the consumption of safe foods. Preference for different NCOs varies. For example, consumers who purchase local products under CSA and U-pick up are likely to attribute high use values to the NCO as they willingly take the trouble to travel to farms to harvest or otherwise help the farmers. In contrast, consumers buying local products from road-side stands or ordinary retail markets may have no strong motivation to protect local agriculture, but may only wish to purchase fresher foods.

The following two cases are a CSA type activity and local government assistance. The former is performed purely under private initiative and the latter is a prototype of public involvement for the *chisan-chishou* movement. These are two extremes in the context of public-private initiatives. There are many other *chisan-chishou* groups practising under public-private partnerships.

The Case of Eniwa (CSA type)

There seem to be several cases of a CSA approach applied in Japan (Sasayama, 1999). One example is found in the city of Eniwa in Hokkaido prefecture (one of the four main islands), which is located about 20 km south from the heart of Sapporo, the capital of Hokkaido. Its steadily growing population was about 66 000 in 2002, among which 1.2 thousand (4.2%) are farmers. Its average farm size is 7.9 ha. The main agricultural products are vegetables, rice, milk, flowers, and tuber crops, but production has fallen gradually as the number of farmers has fallen and farmlands have been transformed into residential or industrial lands.

Eniwa's CSA project began in 1996 through the efforts of a small discussion group created in the 1980's, and whose members were young staff of the Eniwa municipal government who sought to revitalise the city of Eniwa. From the 1990's, they encouraged the participation of citizens in their discussions and held various events. The group found that many citizens were interested in regional agriculture when they held a U-pick up of locally produced melon. This triggered a start of CSA type activities and organised consumers and producers who wished to participate. The current number of participating consumers and farmers is about 50 and 2, respectively.

The principles of this project are that the group consumers must shoulder the risk of crop failure, help with the harvesting of crops, and participate in other agricultural practices as much as possible. The aim is to create mutual understanding between consumers and farmers through dialogue. In this sense, self harvesting is intended to urge members to go to the farm. Moreover, this saves harvesting and delivery costs and reduces the physical burdens for producers, especially elderly ones. When using chemicals, one of the members' biggest concerns relating to food safety, members are fully consulted by the group farmers.

Crops traded in this scheme are limited to potatoes, pumpkins, and water melons. The growers offer shares of each crop to consumers before planting them. For potatoes, the minimum share is 0.25 square meters priced at JPY 5000 (about EUR 37). Harvests are equally allocated to all shareholders, but the quality and quantity produced are highly dependent on the climate each year. For example, the number of shareholders decreased following a poor harvest. About 40 persons have been core members from the start of the project, and another 10 exit, enter, or are replaced. Total membership has been stable in recent years, suggesting that a substantial increase in subscribers is not expected without changes in membership rules. The obligation to help with the harvest as well as risk sharing seems to limit potential members as they should live near the farm. This means that potential markets for Eniwa type CSA is geographically limited. Further, any consumer would want a stable supply of crops.

Aside from this CSA activity, there are other *chisan-chishou* activities in Eniwa. A supermarket contracts with local farmers to market their products, which are specified as being local products. Another activity is the U-pick up.

The Case of Mie (local government assistance)

Many prefecture governments[2] where agriculture is a predominant industry have promoted the *chisan-chishou* movement. In the Mie prefecture, located in the middle of the main island, the prefecture government established an organisation in 2000 to promote *chisan-chishou* by providing consumers with information on *chisan-chishou* voluntary activities performed in Mie and it also provides financial and technical supports for such activities. The organisation has two permanent staff and its operation fund is financed by the prefecture. Despite this, its management is under the control of the administration committee formally de-linked from the government, whose members are representatives from the private sector and individual consumers and farmers.

Chisan-chishou information collected by local government staff or voluntarily provided by individuals is made available to people in Mie through the organisation's website and newsletters. The information includes that of markets selling local products, a list of retailers that cooperate with *chisan-chishou*, the introduction of local specialities for agricultural products, and various events such as workshops and festivals related to *chisan-chishou*.

There are at least twenty small voluntary projects initiated by groups or individuals to promote or implement *chisan-chishou*. The organisation has provided subsidies to about half of the projects. They include cooking classes using local products, a workshop to discuss *chisan-chishou*, a class to teach how to make compost from kitchen waste (recycling of natural resources), and cultivation experiences.

Institutional setting focusing on governmental policies

Chisan-chishou is a local voluntary movement. The key to *chisan-chishou* is how consumers are motivated to purchase local agricultural products, and it seeks to increase sales of the products. Because governments must treat all citizens fairly, they cannot specifically support *chisan-chishou* products to the detriment of others, their role in *chisan-chishou* must be limited as shown in the case of Mie, to activities such as the creation of a network of consumers and suppliers on *chisan-chishou* and the provision of information. Therefore, their involvement is less costly and minimal. The central government is not concerned at all.

Measures to promote NCO transactions

To motivate consumers to take part in *chisan-chishou*, the dissemination of this concept to prospective consumers and the provision of commodity information are important. As the *chisan-chishou* movement is relatively new, it is not yet well-known in Japan. As witnessed in the Mie case, education on the concept of the *chisan-chishou* through workshops and conferences seems to be the first step. For information, its integrated provision by an organiser also seems to be more cost-saving and convenient than are individual producers and retailers. Consumers seeking such information, especially, enjoy greater benefits from this integrated information source. Further, the producers' and retailers' financial capacity does not seem to allow extensive advertising of local products.

Costs in getting the mechanism to work

The costs to start and operate a *chisan-chishou* seem to be minimal. Producers change from wholesale outlets in the city to a local market to sell their products. For this purpose, they diversify the crops grown, but considering the size of local markets, the number of participating farmers is not large. Consumers pay a little more for their food products at the retail level. Expenses for organisations supporting this type of movement are limited to administrative costs. Costs overall are low, because of the voluntary nature of the movement, which entails no legal or administrative obligations.

Assessment of institutional and other arrangements and conclusion

As discussed, *chisan-chishou* is a voluntary movement that has no unified approaches with consumers pursuing their own purposes through the purchase of local products, which includes safe and fresh foods, better rural economy and environment, and the rediscovery of local products and cuisine. The movement seems to be effective for buying safe and fresh foods from local producers working for *chisan-chishou* consumers rather than from other regions, although there is no clear evidence to this. The effects on the rural economy and environment cannot be observed yet. Other than CSAs the consumer's message to the producers relates to the desirability of the continuation of agriculture in their locality but without any additional payment for NCO conservation. This is a weak message compared to other NGAs that provide payment for NCO enhancement and conservation. Unless *chisan-chishou* stimulates local agriculture and expands production, the impact on the local economy and environment will not be discernible. If the movement attracts more consumers, net local production and profitability would increase and the farmers would then perform more environment friendly farming.

Each *chisan-chishou* movement generally confines its participants and benefits to the locality. Therefore, geographical balance and equity across incomes, if they are issues, should be considered within the locality. Under the present situation where *chisan-chishou* market is small and corresponding NCOs provision cannot be observed, issues associated with equity have not emerged.

As long as no payment is made, the NCOs remain unvalued. However, if consumers pay more for local food products that are placed beside the same, but cheaper products produced outside the locality, the price difference would imply a valuation of local NCOs, and/or of the quality and safety aspects thought by consumers to be embodied in them.

Any voluntary activity has a relatively high risk of discontinuation. Consumer behaviour is a dominant force in the evolution of a *chisan-chishou* movement and no cost is required for entry and exit. For example, in case of a recession, consumers may prefer to buy cheaper products and it is possible that *chisan-chishou* products would be less popular. In this sense, *chisan-chishou* arrangements may not be very stable.

Notes

1. Food mileage, originally advocated and promoted by a UK private organisation, Sustain: The Alliance for Better Food and Farming, is an indicator of travel distance of food products from farm to dining room table calculated by the volume of food products (ton) multiplied by the travel distance (km) (Nakata, 2003).

2. Japan's administrative hierarchy under the central government is 47 prefecture governments and 3 000 city/town/village governments.

8. Voluntary flood mitigation (Type P5)

Abstract and classification

Paddy rice growers voluntarily mitigate downstream floods by delaying the rainwater discharge in cases of heavy rain. Such decisions are motivated by a sense of community solidarity and there is no pecuniary remuneration for the farmers' action. This voluntary action can be categorised as Type P5 where there are no transactions costs in the provision of NCOs.

Background

Kanbayashi village, located at the mouth of several rivers in the Niigata prefecture located on Japan's main island, produces high quality rice but has suffered from repeated floods. This is due primarily to its low ground level, 0-2 meters above the sea level. The most flood-prone area in the village was the estuary up until 400 years ago, but this area had been gradually reclaimed up until 40 years ago to meet the growing demand for rice. The most serious flood in recent years occurred in 1969 when 108 houses collapsed and 19 persons were killed. Since then, the village has suffered a total of eight floods.

Non-commodity outputs addressed and the mechanism used

Plots of paddy fields have a flood mitigation function as they can store water. Paddy fields have 30 cm high ridges between plots or ditches, these structures are necessary as rice is grown in water or in very wet conditions. During the growing period, water is carefully supplied or drained depending on rice growth and ground conditions. If rice is submerged after the emergence of rice ears, yield is seriously affected. To avoid this, farmers prevent water from flowing into their plots from upstream and drain rainwater as speedily as possible so as to offset possible flooding. Nonetheless, the plots inevitably store water to some degree due to small outlets relative to plot's area.

A tacit rule for mutual help under a pre-modern irrigation scheme

There used to be a traditional local rule on flood mitigation in Kanbayashi where the average area of the paddy plot was small (about 0.1 hectare) and irrigation water was distributed through curved earth irrigation-cum-drainage ditches. This tradition has disappeared with the renovation of a pre-modern irrigation scheme and associated water management. Under the old regime, in cases of possible floods, farmers closed the entrance mouth of the ridges to prevent the inflow of water from the ditches (the mouth was usually open to drain water). In heavy rains, however, water flowed over the ridges and was stored in the paddy plots. Farmers could do nothing about this, but once the water began to subside, they drained water as early as possible by cutting the ridges to help the rice to emerge from under the water. In this final process of drainage, Kanbayashi farmers had a tacit rule that water be drained first from plots of downstream farms, otherwise these plots had to wait until the upstream farms had completed their drainage process. Depending on the magnitude of the flood, it could take up to a day to complete the drainage of all farmlands in the area. This tacit rule did not involve any documented agreement or a pecuniary compensation to upstream farmers who were exposed to the risk of a reduction in yield. According to a Kanbayashi farmer, upstream farmers were motivated by the fact that they had an advantage in withdrawing water primarily from ditches, a great privilege in times of drought, and that they were less exposed to the risk of floods than were downstream farmers. This sense of mutual help was gradually developed among the farmers who used and maintained the same irrigation scheme.

A new arrangement under a modern irrigation scheme

The renovation of the irrigation scheme and land consolidation, carried out from 1981 to 2002, to raise the work efficiency widened the unit of paddy plots to an average of 0.3 hectare and ditches were realigned — irrigation and drainage ditches are now separated, winding ditches have been straightened, and earth ditches are lined with concrete. However, the speed of rainwater being discharged has increased because of the separation and renovation of straight concrete lining drainage ditches as well as because of the recent increase in the conversion of farmlands to residential or commercial lands. This speedy discharge substantially reduces the risk of flooding in the upper paddy plots, but likely increases the risk at the downstream level. Furthermore, it has destroyed the tacit mutual help rule because water in the upstream plots is automatically flowed out down stream, which in turn discourages them from storing water to help farmers located downstream.

In this situation, farmers and non-farmers living in flood prone areas have organised, with the assistance of local government, a forum to discuss counter-measures against future floods. A proposed arrangement is that a square board, 40 square cm with a hole of 5 cm in diameter (originally set at 15 cm), is set in a concrete drainage box installed at the outlet of every plot; this is expected to delay discharges from paddy plots to drainage ditches and result in storing water. The key principles of the proposed arrangement are that participation of upstream farmer's is voluntary and that the risk of reduction in rice production shouldered by upstream farmers is either zero or minimal. A local government has assisted this arrangement, presently being tested, by providing the boards to participants (at a small cost). In July 2004, torrential rains hit large areas, including Kanbayashi, and, according to a local government officer, it was found that the discharge of rainwater was reduced. There has been no scientific corroboration of this statement however.

The big difference between the old and new arrangements is that the former aimed to smoothly drain flood water when a flood subsided to mitigate damages to the rice. The latter cuts peak flow by delaying water discharge from upstream, thus aiming to lessen the number of areas affected, including houses and various infrastructure. In this regard, the latter is a substantial measure towards stabilising flood water.

Costs and difficulties encountered in getting the mechanism to work

The conventional measures for flood control use river bank improvement, drainage canals and drainage pumping stations. These require huge initial investments and entail significant maintenance and operational costs. Their installation takes time due to government budgetary constraints and/or the construction period. In comparison, the voluntary arrangement is a strategy that requires both less time and less investment.

The effectiveness of the new arrangement is not yet scientifically proven and the full participation of upstream farmers cannot be secured without legal or administrative obligations. Furthermore, the layout and optimal scale of other flood mitigation measures have and will be determined based on the local flood control master plan that has not taken into account the new arrangement despite its potential significance. The arrangement should be built in the plan for future flood control works, but that seems to be difficult without scientific justification and no obligation of upstream farmers. This implies a problem in that voluntary actions inherently exclude any obligations.

Assessment of institutional and other arrangements and conclusions

This case study involves simple institutional and organisational settings, which means that they entail the least transaction costs to set up and maintain. The activities do not require contracts, strict enforcement and monitoring, or pecuniary rewards to farmers. Solidarity among farmers is a key informal institution that motivates upstream farmers to help the downstream dwellers. Its development, however, is based on a gradual collaborative effort for the operation and management of irrigation schemes and other farm exercises.

The new arrangement seems to achieve a high level of efficiency in terms of benefits against costs, although the magnitude of the benefits has not yet been examined. In terms of equity, all persons living in flood prone areas benefit equally regardless of their income levels and the demand for flood mitigation is a high priority for all persons. For a geographical balance, because the arrangement is well targeted to improve the flood mitigation capacity of paddy fields, there are no regional imbalances. High sustainability is also expected in view of minimum maintenance cost. Indeed, maintenance costs will involve primarily the occasional placement of the boards. In order to sustain this new arrangement, it is important to minimise the costs and risks shouldered by upstream farmers.

A notable factor of this case is that beneficiaries of flood mitigation and its contributors are living in the same communities. In addition, the beneficiaries can be easily identified. This closeness between beneficiaries and NCO suppliers contributes to creating a sense of solidarity and produces the basis for a low probability of disputes. However, it could be difficult to find areas that have such geographical features as found in Japan, although similar arrangements could be made elsewhere on a smaller scale.

9. Voluntary conservation of biodiversity and landscapes on Banks Peninsula, New Zealand (Type P5)[1]

Abstract and classification

A community-based trust works with landowners to voluntarily protect biodiversity and landscape through conservation covenants and related activities. It has been effective in promoting vegetation conservation, but questions remain about whether it can successfully manage the wider public interest in landscape character and function. The regime is classified as a Type P5.

Background

New Zealand undertook dramatic economic reform in the 1980's (Holland and Boston, 1990). Government support for agriculture was removed, and farmers became fully exposed to the global commodity market. This forced them to become more independent from government, and created a distrust of governmental organisations. In 1991, the planning system was also reformed to create an "effects" based approach (Memon and Perkins, 2003). This established a public interest in non commodity outputs of agriculture such as biodiversity and landscape. Voluntarist methods of management were strongly encouraged where appropriate,[2] but in practice most planning authorities continued with a largely regulatory approach.

One unintended consequence was that "landscape protection" on agricultural land became highly contested during the 1990's. The national farmers' organisation (Federated Farmers, www.fedfarm.org.nz) became a vocal critic of planning authorities that implemented landscape and biodiversity policies, and several withdrew proposed plans based upon regulatory regimes (Peart, 2004). The case study examines an attempt to reduce such conflicts through introduction of a local voluntarist regime to manage non commodity agricultural outputs.

Banks Peninsula/Horomako[3]

Banks Peninsula is the dissected remnant of two extinct volcanoes on the NZ South Island East Coast, with an indented coastline and visually distinctive ridges and rock features. Only 1% of indigenous forest cover remains (Wilson, 1992), replaced by exotic grasses and trees associated with pastoral agriculture. Introduced pests and weeds threaten the remaining indigenous biodiversity. Nonetheless the landscape is highly valued, with a mix of natural and cultural features, and a significant area (15%) of regenerating forest (Wilson, 1992). Whilst pastoral farming has become marginal in many parts of the Peninsula, land values have risen as part of a global real estate market seeking "lifestyle" properties associated with the picturesque and frequently spectacular landscape (Primdahl and Swaffield, 2004).

The Resource Management Act 1991 (RMA)

The RMA promotes "the sustainable management of natural and physical resources". This includes, *inter alia*, as "a matter of national importance", a requirement to preserve the natural character of the coastal environment, protect "outstanding" natural features and landscapes from inappropriate subdivision, use and development, and protect areas of significant indigenous vegetation and significant habitats of indigenous fauna.[4] The RMA is implemented at a local level through District Plans.

The Banks Peninsula Proposed District Plan (BPPDP) was released for public comment in 1997. It specified a range of types of "protected" area, to be managed for landscape, biological, landform or coastal values, and proposed conventional regulatory mechanisms, including a maximum density for new rural housing of one per 20 ha. The regime attracted criticism from rural land owners, and threat of legal action from Federated Farmers. The District Council established a Rural Task Force to address the concerns. They proposed a revised approach based largely upon voluntary methods of environmental protection on privately owned rural land.

In 2001 local landowners established the Banks Peninsula Conservation Trust (BPCT) to implement such a regime, and in late 2002 the Council publicly notified a variation to the 1997 Proposed District Plan. This introduced a liberal approach to regulation of rural activities and included a new objective, "To encourage the development of non-regulatory methods of achieving the conservation and enhancement of the valued landscapes, vegetation and wild life of the Banks Peninsula" (BPDP P182). Variation No2 significantly reduced the number and extent of "protected" areas in the District. It also introduced a more permissive approach to rural subdivision, providing for consent for a house site on a rural land title of only 2 ha, if it is linked to an adjacent new conservation covenant. The formation of the BPCT was an important influence upon the adoption of the Variation.

The Banks Peninsula Conservation Trust (BPCT)

The BPCT is a not-for-profit trust with a Board of Trustees and a Management Committee, mostly local farmers and/or landowners. Its vision is "to create an environment in which the community value, protect and care for the biodiversity, landscapes and character of Banks Peninsula".[5] The Trust Deed itself, however, does not refer to landscape, and states its aim as "to promote sustainable management and conservation on Banks Peninsula" (BPCT Trust Deed, 2001).

The BPCT takes a proactive approach to conservation, by approaching farmers to ask if they are willing to protect areas of land on their property that have been identified as conservation priorities.[6] This is achieved through a covenant agreement to maintain and enhance the indigenous vegetation, usually by fencing out grazing animals. The BPCT offers some contribution to fencing, surveying and registration costs, and helps farmers apply for funding from other sources, although full costs are seldom if ever covered.

There are two types of covenant, one in perpetuity, the other for 20 or 25 years, recorded against the land title. Land remains in private ownership, with no obligation to provide any public access. Trust members also collectively undertake practical conservation activities (*e.g.* pest eradication), provide educational material and events (*e.g.* workshops), and undertake conservation advocacy (*e.g.* submissions to Council). One of the primary motivations for farmer involvement appears to have been such peer support, which reinforces individual commitment to conservation outcomes.

Funding for the operation of the BPCT itself comes primarily from central government (BPCT Annual Report 2004). The Banks Peninsula District Council has identified the Trust as a key local organisation through which it expects the statutory objectives and policies of its District Plan to be achieved. It has adopted a policy to provide for 100% remission of rates (local government taxation) upon covenanted land, although this is not yet operational and does not appear to be a major incentive to farmers at present.

Evaluation and discussion

The BPCT has attracted recognition within New Zealand as an exemplar of a voluntarist approach. In 2003 it received a premier "Green Ribbon Award" from government for "outstanding contribution to sustaining, protecting and enhancing the New Zealand environment".[7]

By early 2005 the BPCT had registered five covenants totalling 48 ha., with 14 proposed covenants in progress, and a further 15 in "the early stages".[8] Areas range from 1ha to 75ha, mostly existing bush, plus several distinctive landforms. This can be compared with the effects of another voluntary regime for biodiversity protection and enhancement. The QE II National Trust registers "open space" covenants on rural land for conservation purposes throughout New Zealand. Between 1985 and 2004 the QEII registered 42 covenants on the Banks Peninsula with a total protected area of approx 570 ha, in sizes ranging from under 1 ha to 77 ha. In its first four years of operation, therefore, the BPCT initiated a significant number of covenants, but the total area protected at this stage is modest compared with pre-existing provisions.

The wider landscape outcomes of the voluntarist regime are less certain. In New Zealand, the presence of structures- particularly new building- is widely believed to detract from landscape value of rural areas (Swaffield and Fairweather, 2003). However the character of such areas attracts people to want to live there, and this creates a major tension. The (liberal) BPDP Variation No 2 came into effect immediately it was notified and was followed by a significant number of applications for resource consent for subdivision and building in rural areas. Approximately one hundred and ten consents have been approved in the Rural Zone since its introduction.[9] There is anecdotal evidence of an increase in the development of "lifestyle" housing in the rural area, and some farmers acknowledge that the more liberal approach to rural subdivision has brought them economic benefit, while regretting the wider effects upon the viability of the rural agricultural system (Primdahl and Swaffield, 2003).

In 2004 the Auckland based Environmental Defense Society published a case study examination of landscape protection policies on Banks Peninsula (Peart, 2004). It noted that the "main pressures on the landscape in Banks Peninsula are the erection of dwellings, plantation forestry and marine farming". The report concluded: "voluntary methods have to date focused on biodiversity conservation, and have yet to grapple with landscape protection issues which may prove more complex. In the absence of restrictive district plan provisions or effective voluntary measures, the district's landscapes are vulnerable to increasing development pressures".[10]

The BPCT 2004 Annual Report included the following statement: "the [management] committee had a workshop on the position that the Trust will take on landscape matters, as increasing pressure comes to be actively involved in the protection of landscapes on the Peninsula. After considerable heartache, the committee has agreed that while the Trust is happy to register open space landscape covenants against land titles, it will remain neutral on the politics, as it was felt that the Trust needs to retain the confidence and support of landowners, agencies, and the more conservation minded members of our society." In other words, while the local farming community has been successful in the development and introduction of a voluntarism regime for protection of remnant vegetation, it has been unable to find ways to manage landscape change that arises from development pressure under the new liberal regime.

Conclusion

The case study shows how a voluntarist regime for some non commodity landscape outputs such as biodiversity and vegetation conservation can be introduced successfully "from the bottom up". There are also social outputs for the farming community through strengthened rural networks. However the acknowledged problems faced in dealing with the effects of development on landscape suggest that voluntarist regimes have difficulty in managing a wider public interest in landscape character and function. The existence of the BPCT has allowed the District Council to step back significantly from the political challenge of how to manage and conserve overall landscape character in the rural area. At present, however, by its own admission the BPCT does not address the effects of development on landscape, and appears unlikely to do so in the foreseeable future. The management of long term multifunctionality of the agricultural landscape of the Banks Peninsula remains uncertain (Primdahl and Swaffield, 2004).

Notes

1. This case study was prepared by Simon Swaffield, Lincoln University, New Zealand.

2. http://www.mfe.govt.nz/publications/ser/ser1997/chapter10.pdf *The State of New Zealand's Environment.* http://www.mfe.govt.nz/publications/biodiversity/biodiversity-outside-public-lands-update-oct01.pdf *Biodiversity outside public conservation lands: An update on Government progress.*

3. Horomako is the name used by the indigenous culture, Maori.

4. Section 6, subsections 1, 2, and 3.

5. BPCT Vision Document:- http://www.biodiversity.org.nz/policies/files/Extranet.

6. Priorities are based upon an inventory of potential sites published in the 1992 Banks Ecological Region survey report.

7. New Zealand Ministry for Environment website.

8. K. Townsend, Conservation Projects Officer, BPCT, Personal Communication, 23 February 2005.

9. B. Hoffman, BPDC, Personal Communication, 22 February 2005.

10. Peart (2004), page 79.

10. Direct transaction: a case of mineral water in France (Type N1)[1]

Abstract and classification

This case study is categorized in Type N1 in that it shows a private solution to a conflict over agricultural nitrate pollution. The polluters are farmers and the pollutee a mineral water bottler. Key factors for the solution involved clear property rights, easy identification of stakeholders, strong involvement of researchers, and payment from pollutee to polluters.

Background

Nestlé Waters[2] is the world leader in bottled water and established in 130 countries. The bottler includes several famous brands such as Vittel and Perrier. In the early 1970's, intensification of farming practices in the Vittel area, located in the French Vosges mountains, led to concerns about imbalances in the local ecosystem (Nestlé, 2003). More precisely, in 1988, the Vittel production unit[3] noticed a quality deterioration of its mineral water, notably a slow but regular and significant nitrates increase. The main cause was non-point source pollution from intensive farming practiced in the fields surrounding the Vittel springs, the so called "small Parisian basin". These upstream farmers (about 40 farmers for 3 500 ha) are mainly milk and cereals producers. Dairy production is based on corn, which is considered an important factor in the increase in the use of nitrates (Deffontaines and Brossier, 2000; Perrot-Maître and Davis, 2001).

This issue was critical for Vittel. Indeed, while European regulation limits the maximum level of nitrates to 15 mg/l for mineral water for infant feeding, some countries impose a tougher threshold nitrate level of 10 mg/l (INRA, 1997). Vittel attempted several unsuccessful strategies to deal with this problem such as the use of regulatory pressures, collaboration with the Chamber of Agriculture, purchase of fields and meetings with the Ministry of the Environment concerning significant changes in farming practices[4] (INRA, 1997). Furthermore, the reported alternative of moving to new springs was not accurate because Vittel would lose the reputation gained from the asset tied to a specific and famous location.[5] According to Perrot-Maître and Davis (2001), Vittel "has come to realise that protection of water sources is more cost effective than building filtration plants or moving continuously to new sources."

Therefore, Vittel turned its attention to research by contracting with the French National Agronomic Institute (INRA) for a specific research-action program, the so-called AGREV.[6] Indeed, researchers from INRA were familiar with the local agriculture because of previous collaboration with farmers on agricultural development issues (INRA, 1997). The question of Vittel was expressed as follows: "What changes are required concerning farming activity, used on the site, and under what conditions in order to reduce the rate of nitrates found beneath the roots of cultivated plants and grassland, and to ensure that this rate remains below the limit of 10 mg per liter?" (Deffontaines and Brossier, 2000). This question initiates the negotiated management program described below.

A description of the negative externality addressed in the case study and of the mechanism used

The negative externality addressed in the Vittel case is the deterioration of the water quality caused by intensive farming practices. The percolation of nitrogen runoff and intensive pesticides use affect the quality of the bottled water. Vittel bought several fields (about 1500 ha), *i.e.* acquired property and tenant rights close to its springs at very attractive prices (Chia and Raulet, 1994; Brossier and Gafsi, 1997) and became the owner of 45% of the sensitive area shaping the water quality.

The mechanism used to address this negative externality is a direct negotiation between the pollutee and polluters. This solution was feasible because the involved parties were easily identified, not too numerous and the definition of accurate property rights possible at a reasonable cost. The contracts allow (1) the definition of property rights on production choice and agricultural practices and (2) the transfer of property rights from the formal owners or users of the fields, *i.e.* the farmers to the industrial company, Vittel. Formally the contractual arrangement is almost a purely private agreement to internalise some negative externalities of intensive agriculture. In other words, farmers agreed to switch to less intensive dairy farming and pasture management. The property rights transfer lasts for a limited period and farmers are rewarded in several ways such as income support, equipment subsidies, and free technical assistance. The payment is not based on the improvement of water quality but rather compensates the farmer for the risk and reduced profitability associated with the change to a new agricultural system (Perrot-Maître and Davis, 2001). From the Vittel point of view, this arrangement can be considered as more cost-effective and sustainable than a permanent compensation. Note that the fields previously acquired have been used as powerful incentives to encourage farmers to accept the contractual arrangement by making these fields available to them under contract. According to the study by Gafsi (1999) on a sample of farms, these farms have increased their average usable agricultural surface by 34%. These fields are exploited by farmers under contract, but remain under the control of Vittel (Gafsi, 1999). Despite the initial reluctance of some farmers, the number of farmers under contract has grown and reached a rate of 92% of targeted farmers (Barbier, 1997; Gafsi, 1999).

Institutional settings focusing on governmental policies

A close examination of the situation shows that the problem cannot be solved by enforcement of existing laws and regulations, such as the Water Act of 1964 (INRA, 1997). Consequently, private agreements between the two parties appear as the most cost-effective solution. Formally, the contracts are traditional contracts between Vittel and farmers defining the use of some rights. No formal or direct governmental intervention was necessary to define, implement or enforce these private agreements. The only indirect and significant governmental intervention was the strong implication of an interdisciplinary research team including scientists from several public research agencies. This research team played a strong role in defining precisely the rights which have to be included in the contracts to achieve the desired performance in terms of water quality. Moreover, the research team played a role of mediation and improving mutual comprehension between *a priori* divergent and asymmetric interests of the two parties, *i.e.* an important industrial company trying to improve its water quality and farmers aware of public concerns but fearing the change of their production systems. Of course, the public authorities played a fundamental role in providing a credible legal system to assure the enforceability of contracts and granting some limited financial aid (Perrot-Maître and Davis, 2001).

Organisational arrangements

The organisational arrangement is a private contractual one. The research team helped to define the specifications and clauses of the contracts and the obligations of each party. To contract with targeted farmers, Vittel negotiated with each of them and proposed individual incentives and compensations. The incentives provided by Vittel to encourage farmers' acceptance were variable among farmers according to their individual situations. For example, the geographic proximity with Vittel springs was a strategic variable in the bargaining phase. The main (and average) obligations of the farmers and Vittel are described in Table 3.

Table II.3. Main obligations of farmers and Vittel

Farmers	Vittel
Eliminate corn crop	EUR 230 per ha and per year during 7 years
Ban pesticides	Equipment investment of about 150 000 euros per farm (haymaking materials, barn drying, buildings, etc.)
Compost all animal waste	
Nitrogen fertilisation by composted manure (an additional nitrogen contribution less than 30 units per ha is tolerated)	Free supplying of manure treatments and use (composting, spreading, etc.)
	Free technical assistance
Limit one livestock unit per ha of grazing area and balance livestock feed	Free usufruct of the previously bought fields and the quotas associated (about 25% more)
Ensure farm buildings are up to Agrivair standards, exceeding legal obligations	

1. Farmers have substituted corn by Lucerne and lost the Common Agricultural Policy (CAP) aid attributed to this crop.

2. The services supplied by Agrivair represent 23% of the overall seasonal works for each farm (Gafsi, 1999).

Main Source: Gafsi (1999), Nestlé (2003).

As mentioned above, the payment is not indexed on the improvement of water quality, but based on the switching costs and compensations resulting from the adoption of a less intensive farming system. To ensure its obligations and prove its sustainable implication in the radical change, Vittel has created an agricultural advisory firm, Agrivair. The mission of this firm is to advise, accompany, manage and enforce contracts with farmers (Gafsi, 1999). An interesting feature of the enforcement is the use of scientific research procedures that have been adapted for other purposes than their initial use (Chia and Raulet, 1994).

In terms of performances, the records show that the overall nitrates rate in groundwater has decreased. Fifty per cent of monitored springs experienced a decrease of the nitrates rate and the other 50% have a constant nitrates rate (Gafsi, 1999).

Other measures that have facilitated the transaction

As briefly mentioned before, the strong involvement of INRA and other researchers has facilitated the design and implementation of the project. The financial participation of the French Water Agency in the research program was also significant (INRA, 1997). Whilst the desired outcome by Vittel was clear, the ways to achieve it needed to be defined. The interdisciplinary research team played an important role in building bridges between the end result expected by Vittel and the farmers' management practices. Rather than "ready to use" solutions, the research team, with the assistance of the farmers, progressively elaborated technical and economically feasible solutions compatible with farmers' strategies. This collaborative process contributed to increase farmers' acceptance because the clause of the proposed contracts was co-built and integrated farmers' concerns (INRA, 1997; Gafsi, 1999). During the process of applied research, an extension specialist was recruited to ensure constant communication between farmers, Vittel and the research team (Gafsi, 1999). It should be noted that the person in charge of Agrivair was the same agricultural extension specialist previously recruited by the research team (Deffontaines and Brossier, 2000; Chia and Raulet, 1994). Agrivair has recently introduced new technologies such as a information geographic information system to manage sewage spreads, which can increase the quality of its services. Several clauses of the contracts relate to the prevention of fraud, such as free access to accounting documents and visual inspection of farms. According to Chia (2004), "visual inspection is sufficient and very easy for anyone experienced in agriculture." Finally, several farmers of the Vittel perimeter have switched to organic production, allowing a better valorisation of their agricultural products (Reibel, 1999).

Costs and difficulties encountered in getting the mechanism to work

Vittel has incurred different costs in getting the mechanism to work. We distinguish three kinds of costs: (1) the *design costs* including the contract with the INRA and other costs for defining the accurate area to buy or put under contract, the property rights to contract, the terms of contracts with farmers, (2) *implementation costs* notably including buying fields and investments in individual farms under contract, the costs associated with creating and running Agrivair, and (3) *enforcement costs* such as the economic compensations negotiated with farmers for changes in farming methods and the costs of accompanying and monitoring farmers.

Although Vittel was a major employer in the small region, it had little knowledge of the farmers' realities and reasoning (Barbier, 1997). For example, the regulatory context was perceived differently by Vittel and farmers. Indeed, farmers were arguing that drinking water requires a maximum of 50 mg nitrates per liter which was achieved, but mineral water must satisfy a maximum of 15 mg per liter threshold. Moreover, another important difficulty of the project was the negative reaction of the local agricultural bodies, such as the Chamber of Agriculture, which felt threatened by an intrusion of researchers in their field of competence. This tension between traditional extension services and the research team put farmers in an uncomfortable position (Barbier, 1997; Chia and Raulet, 1994). Some farmers have been reluctant to accept the proposed contracts because, for example, of union activism.

For farmers, this radical change corresponds to the learning of a double competency, *i.e.* low input agriculture and water protection. Moreover, the farmers under contract suffered from media pressures and jealousy, notably from other farmers not included in the Vittel perimeter. Conflictual relations occurred at times between Agrivair[7] and farmers (Reibel, 1999). A potential obstacle to the transposition of such a model relates to the technical and financial dependence of farms with regard to the other party involved in the contract (Brossier and Gafsi, 1997). The results in terms of nitrates decrease are not immediate. Moreover, the process takes time and is very progressive to allow the farmers' to learn the new practices and understanding their contractual relationships with Vittel.

Assessment of the arrangement and conclusion

The approach has been applied to other companies, *i.e.* Perrier and Contrexeville. The application of the same approach to the Contrexeville springs was relatively easy because of the geographic proximity. According to Perrot-Maître and Davis (2001), "the Perrier springs are located in southern France in an area of vineyards and intensive wheat cultivation where phosphates and herbicides are the main sources of water pollution. Perrier successfully introduced organic agriculture to 20 farms that cultivate approximately 350 hectares of cereals and 200 ha of vineyards and regularly monitors over 900 ha of land. The highly favorable market conditions for organic produce made significant contributions to the rapid adoption of improved farming practices around the Perrier springs. Other French bottlers — Evian and Volvic — have considered using Vittel's experience as a model." The tools and approach developed by the research team have also been applied to other cases of water contamination by farming practices, *e.g.* in the Migennois and Plateau Lorrain (INRA, 1997).

Several other issues can affect the potential for transferability (Perrot-Maître and Davis, 2001).

- **Scale**: "The Vittel model may be difficult to use in larger geographic areas or in an area with a greater number of farmers." Indeed, the greater the number of parties, the higher the transaction costs associated with designing, implementing and enforcing an agreement. If the transaction costs exceed a certain level, they can make other alternatives more cost-effective.

- **Timing**: If quality drinking water is needed immediately, the approach adopted by Vittel may be too slow and make the use of filtration plants unavoidable. The timing includes the time needed to design solutions, encourage adoption and the lag between adoption and the change in performance. For example, between February 1993 and February 1996, the proportion of farmers under contract evolved from 3% to 65% and 92% in 1998 (Barbier, 1997; Gafsi, 1999). This stresses the need to consider pollution problems at an early stage rather than when pollution thresholds are exceeded.

- **Private sector profitability**: Given the high level of investment required, imitating the Vittel approach seems limited to highly profitable industries (Gafsi, 1999). The purchase of property rights (land acquisition, practices changes) was possible because the value of the water quality was significantly higher for the bottler than the loss incurred by farmers. The opportunity cost of farmers to accept the contract was lower than the opportunity cost of the bottler. The creation of Agrivair was essential because it was perceived by farmers as a signal that Vittel was really investing for agriculture and that farming changes would really benefit from a long term support (Barbier, 1997). Despite the significant cost of the Vittel approach, it can be considered as a reasonable alternative by taking into account (1) the effects of a *laissez-faire* choice on the image of Vittel water and therefore the expected loss of future profits, (2) the impossibility to move to new sources without bearing the associated costs (*e.g.* search and acquisition costs) and losing the reputation asset associated with the brand Vittel, and (3) the simulation of the research team proving that the program remains profitable when it is compared with another alternative based on a complete switch to well-managed pastures. The switch of several farmers to organic production may have contributed to making the new system more sustainable and profitable (Nestlé, 2003).

- **Strong involvement of researchers**: The multidisciplinary research and extension team played an essential role in the success of the operation. "The research program was finalised in 1996. Seven years of research enabled a preliminary conclusion to be drawn regarding three main aspects. Firstly, the objective regarding sustainable development on the Vittel plateau was achieved. The agrarian system used on the Vittel site has clearly made progress in terms of nitrates rate in the water sources and in terms of farmer incomes. Secondly, knowledge has been produced over a wider spectrum, such as in technical and socio-economic fields. The apparent high cost of the operation does not make this experience prohibitive. If and when drinking water becomes scarce, financial backing could easily be found. The third aspect concerns the positioning of the research team faced with a complex question: the research team – placed in a highly uncertain context having accepted the challenge of a complex question – formulated, set up and implemented a wide range of technical and social tools which brought the various actors together on several levels. This is a good example of negotiated management" (Brossier and Gafsi, 2000).

In conclusion, this direct transaction, based on individual contracts, constitutes a private solution for externality problems where clear property rights and easy identification of stakeholders played a key role. The applied principle was not "the polluter pays", but the counterintuitive "pollutee pays". This arrangement constitutes an original application of Coase's recommendations (Coase, 1960) and transposing it to other institutional environments constitutes a challenging issue.

Notes

1. This case study was prepared by Gilles Grolleau of UMR INRA-ENESAD.

2. Nestlé Waters was previously known as Perrier-Vittel, which was itself known before as the *Société générale des eaux minerales* de Vittel.

3. Hereafter, Vittel designates the Vittel Company, regardless of its formal name.

4. Vittel attempted in 1988 to propose "ready to use" solutions, elaborated by the French Committee for the Reduction of Water Pollution by Nitrates (CORPEN). This solution was to transform all the fields of the perimeter to grasslands. Vittel could buy all the fields and re-allocate them to farmers. The success of such a strategy was limited because farmers were rejecting this solution as not well adapted to their production system. Because of a high level of refusals, Vittel redefined its strategy with the help of the research team (Gafsi, 1999).

5. Note that in France mineral natural water must come from the same springs, while natural spring water can come from different springs, regardless of their respective locations.

6. AGREV is the French acronym for *Agriculture Environnement Vittel*. Note that the *Agence de l'Eau Rhin Meuse* (French Water Agency) was also a co-contractant with the INRA. The financial resources of the research program came respectively from Vittel S.A. (33%), the *Agence de l'Eau Rhin-Meuse* (17%) and the INRA (50%) (INRA, 1996).

7. According to Nestlé (2003), "Agrivair's activities now extend beyond this single farming issue. They are actively involved in managing the forest as well as green parks and golf around the Vittel and Contrex springs."

11. Voluntary approaches (Type N2)

Abstract and classification

Farmers organise groups to mitigate local environmental problems in a voluntary way. Governments, influence their operations by providing financial or technical assistance or policies other than regulations. The salient feature of this Type N2 study case is that farmers' manage the scheme without top-down government regulations.

Background

In 1998, the OECD implemented a study examining farmer-led voluntary approaches to improve the local environment and pursue sustainable agriculture with four country cases.[1] The purpose of the study was to seek the following: a better understanding of the nature, scope, and activities of farmer-led groups; to examine the potential role of the groups in implementing and perhaps even defining the sustainable agriculture agenda; and measures to encourage the formation and development of these groups. Its scope was thus wide with respect to dealing with sustainable agriculture as it was not confined to the reduction of negative externalities.

The objectives and approach of the 1998 OECD study was similar to the present study although the scope was wider and the examination was not always conducted from the same angle. This therefore extracts the key findings and lessons of the 1998 study with limited updating. The focus of in examining voluntary approaches are: the mechanisms used (institutional settings), motivation for the initiatives taken, government involvement, the evaluation of their effectiveness, and suggestions to encourage voluntary approaches.

Introduction and Mechanisms used

Four cases were investigated: landcare in Australia, conservation clubs in Canada, environmental co-operatives in the Netherlands, and various farmer-led activities in New Zealand. They did not necessarily deal exclusively with negative externalities. For example, land conservation in the Australian case included nature conservation. They, of course, differed in their objectives, approaches, and institutional settings.

Landcare in Australia. Australia's landcare movement, characterised by voluntary farmer actions, began in earnest in 1986 to tackle a wide range of land degradation issues, such as dry land salinity and erosion. About 4 000 groups are operational across the nation, and they manage 60% of the land and 70% of the nation's diverted water (Department of Agriculture, Fisheries and Forestry of Australian Government's website, 2004) The groups' activities include the creation and sharing of new knowledge (*e.g.* though workshops or the demonstration of new technologies), the sharing of equipment (*e.g.* tree planting machine), and planning development (*e.g.* to reduce the amount of salt and silt, or to develop business plans). They are funded by national or state governments, private firms, research institutes such as universities, and/or by the farmers themselves. The groups are not composed exclusively of farmers but open to all community members. They elect their own officers and decide what issues to address and how to approach them. The Commonwealth, State and Territories governments are all supportive of their activities and provide technical and financial help, which encourages group formation and operation. The Commonwealth's National Landcare Program in particular provides substantial strategic support to the landcare movement. The program has three components: national, commonwealth/state/territory, and community, each of which has different implementation bodies and scope. The community component provides grants to group activities that contribute to problem identification and solutions through community activities.

Conservation clubs in Canada. Owing to cold winters, the need for insecticides and herbicides are not as great as in some other OECD countries. Nutrient pollution, wind and water erosion are Canada's important agri-environmental problems. To voluntarily mitigate these problems, farmers began forming rural conservation clubs in the provinces of Ontario, Prince Edward Island, and Quebec in the early 1990's, encouraged by the federal-provincial funding available under Canada's Green Plan. In 1991, four farmer groups in Ontario formed an organisation, the Ontario Farm Environmental Coalition (OFEC), to help farmers committed to protecting their lands. The OFEC assists farmers in preparing environmental farm plans and their implementation through workshops. The federal and provincial government provide the operational funds. The development of environmental farm plans is reviewed by a peer committee or by professionals.

Environmental cooperatives in the Netherlands. The first farmer-led co-operative that integrated protection of the environment emerged in 1992. Their purpose and activities are diverse. For example, the Eastermar's Landau Association, which covers a 1 900 hectare area located in the northern part of the country, seeks to maintain deteriorated small canals and earthen dikes covered by low trees that provide wildlife habitats. Its main target is thus the maintenance of NCO, although the improvement of water quality, which is the reduction of NE, is conducted. Much of the work previously undertaken by individual farmers is now done collectively, attaining higher efficiency and level of participation. Among the benefits is the establishment of a network among farmers, various levels of government and researchers, which helps them to access new technologies and to obtain technical and financial assistance. Another group, called the Working Group Cultivation in the Ground, focuses on environmental problems relating to glasshouse horticulture across the country. They prepare an action plan that includes requests for government actions, such as finding an alternative to methyl bromide, a soil fumigant. The emergence of these groups has proved a useful vehicle for mobilising farmer commitment to environmental protection, and for finding ways to shift more responsibility over the implementation of environmental policy to local communities.

Farmer-led environmental groups in New Zealand. In New Zealand, under the Resource Management Act 1991, most environmental responsibilities are assigned at the local level. The regional programs include soil conversion activities, water quality monitoring and control, and pest management. In the early 1990s, 60 farmer-based community groups were formed to address issues connected with sustainable agriculture, some of which receive administrative or financial support, or both, from a regional council or external sources; others are entirely independent. The formation of these groups is spurred by the Rabbit and Land Management Program, a five-year program (1989-1995) designed to tackle the problems related to the sustainability of the rabbit-prone tussock grassland areas through farmer-led activities. In drawing lessons from the program, Elliot and Anderson (1995) stress the importance of not only physical but also socio-economic and institutional factors, the need for decision support system and information to influence changes in farming practice, community-based stakeholder forums to build consensus, strengthen trust, and to sharing information learned. There are highly diversified farmer-led activities, such as Project FARMER[2] that develops computer-based dynamic modelling tools for sustainable agricultural operations; landcare groups that develop practical approaches to addressing land and water problems; and two research institutes that assist farm management practices.

The voluntary farm community groups examined in the four countries focus on local rather than global issues. They were initially less concerned about water than soil quality despite the fact that the issues of nutrient pollution of groundwater are given priority by the regulators. After several years' operation, they expanded their scope to include sustainable land management from integrated, system perspectives, which is encouraged by government efforts. In every case, the emergence of voluntary farm groups appeared to have been motivated by concern about declining farm profitability, increasing awareness of the linkage between ecological and financial sustainability, and the expectation of achieving more satisfactory and acceptable results through their own initiatives rather than waiting for government solutions to problems of sustainability in general, and pollution in particular.

There is a great diversity in the initiatives taken, but the most common activity is to work together in preparing farm plans as well as in the exchange of information. Generally, these farm plans employ a "whole farm approach" to consider all the environmental, economic and sociological factors that can contribute to sustainability. Often, outside advisers helped farmers to prepare their plans. At that time, a number of demonstrations were given to show how to lower costs at farm level. Many farmers worked collectively on specific projects, such as constructing nature inventories, repairing and maintaining earthworks, or eradicating noxious weeds. The community participation rate is not always very high, however.

Institutional settings focusing on governmental policies

Government policies that promote and support the farm-led groups have a direct impact on the formation and operation of the groups. In Australia, both the federal and state governments are actively involved in providing seed money and strategic support to their landcare groups, and policies toward them are well-developed. For other countries, no development of such formal policies was found. All countries provide some support to farmer-led environmental groups, either through direct aid or advice, and usually both. New Zealand sits at one end of the spectrum. Its central government allocates no funds to specifically support landcare-type activities; most funds come from local governments, industry, charitable organisations and farmers themselves. Central governments for the other three countries provide funding to serve more as a catalyst to private spending. In Australia, no funds are provided directly to individual farmers, only to community landcare groups, which can reduce administrative costs. Another policy aspect to note is the degree of the farmer's commitment, which is generally strongly related to his legal commitments. In Australia, Canada, and New Zealand, commitments made by farm community groups to the responsible environmental and agricultural authorities tend to be informal. That is, the commitments are expressed in general terms and do not contain any sanctions in case of non-attainment. In contrast, commitments made in the Netherlands are of a more formal nature. The agreement between farmer groups and the government includes evaluation criteria and, sanctions for non-attainment of goals.

Assessment of the approach and measures to promote it

At the time of the 1998 study, each of the four countries examined had only begun to evaluate the programs or little had been done due to insufficient time to dedicate to sustainable resource use under the farmer-led approach. Therefore, relevant data was not available and the effectiveness of the approaches could not be clearly demonstrated. As a result, lessons learned from the 1998 study were not necessarily complete. Nonetheless, the following points emerged that, and seem comprehensive enough to consider future development of the approaches.

- *The impact on the environment and the sustainable use of resources*. This is the most important target for the activities. Data was not available, however, for all countries. Information derived from farmers' perceptions was only available in Australia, and this was a mixed evaluation. Generally, no impact on soil degradation issues, *i.e.* the spread of weeds and a decline in rangeland productivity, was observed by at least half of the farmers interviewed, while wind erosion was perceived by the majority as either stable or decreasing in severity.

- *Awareness of agri-environmental issues and provision of relevant knowledge and skills*. These are perhaps the most tangible accomplishments as they are the first results obtained through activities such as workshops. Farmers who voluntarily participate in the activities could be more proactive in local environmental problems and indeed already have a high awareness. Most countries mention this as the first benefit of the voluntary approach.

- *Provision of a social focus for rural communities*. Most group members are trained to make a whole farm plan that examines not only economic and ecological aspects, but also the social aspects of the issues. This also helps farmers to identify the root causes of these issues, not just the symptoms.

- *Impact on institutional arrangements*. Collaboration with research institutions is commonly observed, which helps farmers to obtain new technologies. Researchers are able to set proper research priorities through such collaboration and attain a better understanding of the current problems that farmers have. Another important change was that a bottom-up approach was recognised as an effective policy formation instrument.

The 1998 OECD study concluded that the community-based farmer environmental groups provide encouragement for approaches to responsible management of agricultural land and water resources and that these can play a useful role in public policy. However, the suitability of the approach may be limited by the motivation of farmers, and the environmental issues they are both willing and able to address. In this context, a list of practical suggestions from a New Zealand case study provided in Annex 1 are helpful as they encourage greater community involvement in land management issues. All of the suggestions are supported by the experiences of the other countries examined in the 1998 study. In addition, the 1998 study also provided other suggestions that are presented in the following annex and which are based on more general observations resulting from the examination of the four country case studies.

Notes

1. The OECD study *Co-operative Approaches to Sustainable Agriculture* was published in 1998.

2. FARMER stands for Farmer Analysis of Research, Management and Environmental Resources.

Annex

The Rabbit and Land Management Program (R&LMP): a model for encouraging community action

In reviewing New Zealand's R&LMP, Elliot and Anderson (1995) set out a generic model for encouraging greater community involvement in land management issues, which in their view captures what the participants in the R&LMP had learned from the experience.

- Evaluate the historic sequence of events (technical, sociological and institutional) that have led to the current problems.

- Establish stakeholder forums and chart the future in terms of new partnerships.

- In order to meet immediate needs and reduce anxiety, focus on the most obvious technical issue or problem first. Only thereafter move into the underlying issues, and then within a system context.

- Foster partnerships between farmers and scientists.

- Develop decision support systems based on information provided by both professional scientists and farmers. Extend the scope of the decision support systems from the farmers to their communities, and then from the communities to the region.

- Undertake experiments and trials to address the knowledge and information gaps identified through the decision support system. Locate willing researchers within communities wherever possible. Validate the data.

- Provide ways to extend learning experiences from individual farms to the community; help to encourage community leadership.

- In sum, do not dictate decisions; rather, empower the locals.

A list of suggestions made by the 1998 OECD study for encouraging greater community involvement

- Reform agricultural policies so as to eliminate conflicting signals.

- Make some project funding available only or principally to groups, not individuals.

- Emphasise training and encourage partnerships between farm communities and scientists.

- Work with stakeholders to develop indicators and other decision support systems that are useful both to land holders and regional planners.

- Create an enabling environment for the devolution of responsibility.

12. Sustainable Winegrowing New Zealand® (Type N2)[1]

Abstract and classification

The Sustainable Winegrowing New Zealand® programme is a voluntary approach that reduces negative environmental effects from wine production in New Zealand.

Background

Winegrapes cover just over 22 000 hectares (2004) in New Zealand, and represent the largest individual horticultural crop in terms of area. The area has increased nearly fourfold since 1990. In the year ended June 2004, 31 million litres of wine were exported, or 56% of total production. New Zealand wine is positioned as a niche product in export markets, aimed at the premium and super premium categories.

The winegrape growing sector in New Zealand has developed a scorecard approach to vineyard environmental management in its Sustainable Winegrowing programme. This was initially trialled in 1996/7 and by June 2004, there were over 400 members, covering 60% of the producing vineyard area.

The initial driver for development of the programme was the industry's need to underpin its marketing line that New Zealand wines are "the riches of a clean green land" (see for example, http://www.nzwine.com/intro/).

Early motivations for growers to join included the potential cost savings (for example, in pest control), and for some, a perception that the programme provides proof to local government bodies of environmental performance. In addition, some wineries require contract grape growers to be members of the programme, and to be committed to sustainability.

Ongoing advantages for members include the ability to use the Sustainable Winegrowing New Zealand® logo, and access to information through a newsletter, regional member meetings and workshops, and membership of regional discussion groups. In addition, databases and software tools have recently been upgraded to add value to the data collected for members, both by providing tools for analysing their own data within and between seasons, and by enabling comparison against regional and national benchmarks. The programme has been extended to include wineries (introduced in 2002).

Negative externalities addressed

The scorecard covers soils and fertilisers, sward[2] and irrigation, and disease and pest management. The following potential negative externalities are addressed:

- Nutrient leaching.
- Soil compaction/loss of structure.
- Soil erosion.
- Externalities potentially associated with pesticide use.
- Loss of biodiversity.
- Noise pollution.

These are addressed by setting limits on fertiliser and pesticide application rates, and by defining desirable, acceptable and unacceptable management practices.

Institutional setting and government facilitation

In general, subsidies are not provided to improve environmental performance in New Zealand. Central and Local Government regulations and rules provide the essential environmental safeguards. The Sustainable Winegrowing programme goes beyond this basic set of rules to develop "best practice".

Initial funding for the programme's development came from grower levies. Further development was facilitated by two small Government grants. Development of the programme to pilot stage was supported by the Sustainable Management Fund administered by the Ministry for the Environment (covering 30% of the project costs). Subsequently, the Ministry of Agriculture and Forestry's Sustainable Farming Fund has assisted the development of the database and software to allow the recording, analysis and benchmarking of the data collected, to add value for growers and to allow the effectiveness of the programme to be monitored (with the Fund covering almost 48% of the project costs).

Organisational arrangements

The Sustainable Winegrowing programme is administered as a stand-alone entity within Winegrowers New Zealand. Its operation is self-funding.

The programme is based on a self-audit score card covering the issues contributing to sustainable winegrape production. Score cards, spray diaries, and fertiliser and irrigation records are collected from all members annually (with an increasing number submitting electronic versions). Regional trends are analysed and reported back to members. External independent audits of individual vineyards are carried out at least every three years. Unacceptable scores result in growers being placed on a provisional register for a year, followed by deregistration if scores remain unacceptable.

Other measures that have facilitated voluntary actions

The programme has been successful in attracting New Zealand's largest wine companies to be members. Further gains in membership are occurring as a result of the development of the database and enhanced reporting to growers.

Costs and difficulties

Development of the programme required considerable research and scientific input, as well as grower co-operation during the pilot phase. The costs to develop the programme were beyond the ability of the industry to carry without assistance, especially as many growers were unconvinced of the benefits in the early stages. Seeding finance provided by external organisations has therefore been critical to the successful development of the programme.

In the early stages, many of the benefits of the programme could also be acquired by non-members (*e.g.* the methods used to achieve a reduction in pest management costs could be observed by non-members and applied on their own vineyards), reducing the incentives to join, and membership was relatively static for a period. However, the extra benefits of the logo and database/reporting system introduced in 2003/04 have increased membership, as these benefits are not appropriable by non-members.

Assessment and conclusion

The programme has been in operation since 1996/97, and the area covered continues to grow. In 1998/99, there were 145 subscribed members, covering half the land area of the wine industry. In 2000, there were 200 members, and by June 2004 there were over 400 members covering 60% of the producing vineyard area. The extra benefits provided by the database and benchmarking service have resulted in a surge in membership. More complete adoption would be desirable, and this may occur if the use of the logo is shown to provide a market advantage. However, the current level of adoption is high for a voluntary programme, especially as there have been no direct payments made to growers to induce them to join the programme, and to date, growers have not received a price premium for grapes produced under the programme.

Efficiency and stability

Sustainable Winegrowing New Zealand® is an efficient method of achieving reductions in environmental externalities compared with compulsory government measures, for the following reasons:

- Since it is self-funding and voluntary, administration costs are minimised compared with an equivalent compulsory government programme, as growers would not join the programme if costs were high.

- Transaction costs are low and falling, due to the increased use of electronic submission of grower data.

- The programme of external auditing ensures that claimed outcomes are actually achieved. Auditors are currently drawn from the kiwifruit industry's auditing team, which has proved to be very cost-effective.

- Compliance by members of the programme is almost 100%. The programme is seen as owned by the winegrape industry, and advantageous for business, whereas in a compulsory government programme, the incentives for avoidance or "lip-service" would be high.

- A government regulatory measure would be unlikely to provide the database and benchmarking service provided by Sustainable Winegrowing New Zealand®.

- Reductions in unacceptable practices, and increases in desirable practices have been clearly demonstrated, *e.g.* there has been a demonstrated shift away from organophosphate and synthetic pyrethroid use towards insect growth regulators and BT amongst members (Manktelow *et al* 2002).

Sustainable Winegrowing New Zealand® is a successful voluntary programme in terms of achieving high adoption and improved environmental outcomes. Elements underpinning the success include:

- The nature of the industry, which is relatively cohesive. Growers and wine makers are willing to work together to maintain the integrity of the New Zealand marketing image.

- Sufficient funding (from the compulsory industry levy and seed finance from external sources) for the development of the programme, until growers became convinced of the benefits of the programme.

- The programme was piloted on operating vineyards to ensure the approach was straight forward and practical to use.

- Programme requirements are underpinned by sound science and/or regulatory or market demands, so are viewed as justified by growers.

- Wine industry control and ownership of the programme improves grower acceptance and adoption.

- The programme is administered by a professional manager, so roles and responsibilities are clearly defined.

- Growers obtain benefits from membership including reduced pesticide costs and access to information.

- Growers are supported in implementing Sustainable Winegrowing New Zealand® by a training programme.

- External auditing ensures that scorecards are accurately completed.

Notes

1. This case study was prepared by the New Zealand Government.

2. The term sward refers to the vegetative ground cover used in vineyards, often a combination of grass species and flowering herbs that attract beneficial insects.

13. Tradeable manure/animal quotas in the Netherlands (Type N3)

Abstract and classification

Tradeable quotas for manure production and animals were introduced in the Netherlands during the 1990s as part of a mix of policy measures to deal with the problem of excess nutrients. Quotas were implemented to place a "cap" on the level of manure nutrients, with trading rules later added to achieve a better environmental distribution of livestock across the country and between animal types, and to ensure that production expansion in occurs in an environmentally better manner. Trading is conducted between private agents (farmers and brokers) with the government participating as the rule maker and referee. This case study is categorised in Type N3.

Background

A rapid expansion and intensification of livestock production occurred in Western Europe during the 1960s and 1970s. This was made possible by the improved used of fertiliser to boost feed production and through the large-scale importation of duty free feedstuff such as tapioca, soy, citrus pulp and maize gluten. This was particularly attractive for Dutch farmers due to their proximity to the port of Rotterdam, resulting in a rapid expansion in animal numbers, especially pigs and poultry, in the eastern and southern regions. Between the early 1960s and the mid-1980s, the number of pigs increased by 10 million (450%) and poultry by 50 million (125%). Consequently, the national surplus of manure increased dramatically, rising to about 16 million tonnes (75 000 tonnes phosphate (P_2O_5)) in 1987.

The first signs that the expansion and intensification of animal production was causing environmental problems began to emerge in the late 1960s. A factor contributing to the environmental problem is the soil and hydrology conditions that exist in different regions of the Netherlands. The clay and peat soils in the north, west and central regions (63% of the agricultural area) where arable and dairy farming dominate are situated below mean sea level, have an artificial drainage system and are influence by seepage. By contrast, the sandy soils in the east and south (27% of agricultural area) are above sea level and are freely draining. Consequently, nitrate concentrations in the groundwater of clay soils are on average less than 50 mg per litre, the standard set by the 1980 EC Directive on Water Quality, but exceed it by a factor of up to five times on sandy soils (Oenema *et al.* 1998).

It took almost 20 years and many reports about phosphorus saturated soils, nitrate leaching, surface water eutrophication and forest dieback due to ammonia from animal manure etc, before enough political pressure finally forced a government policy response. This began in November 1984 with the Interim Law for the Restriction of Pig and Poultry Farms which prohibited the development of new, and limited the expansion of existing, pig and poultry farms. But there were so many gaps in the rules that the number of pigs and poultry increased by 30% in three years. Mineral surpluses continued to increase, forcing the government to intervene in the late 1980s with additional measures including a farm level manure production quota which eventually became tradeable.[1]

Negative externalities addressed

Because of the difficulty in directly targeting environmentally damaging emissions from farming, such as the actual level of nitrate or phosphate leached into the water, the policy response has focused on input variables as proxies. At the outset the phosphate content of manure was the target.[2] Phosphate was chosen over nitrogen (N) because the P cycle is less complex than the N cycle and a relatively constant amount is produced per animal.[3] It was initially hoped that a decrease in nitrogen problems would naturally follow from imposing limits on phosphate, based on the assumption of a fixed N:P ratio of 2:1. Experience showed that this assumption was invalid and consequently nitrogen specific application rates were introduced in 1998.[4]

An important instrument of the policy response was to establish in the late 1980s a limit on the quantity of manure phosphate that could be produced annually on each farm. Rules would later be enacted to allow trading to take place, and this measure is the focus of this review. Other instruments used in the policy mix include regulations on manure application and storage, payments, levies on nutrient surpluses, and extension and advice. Some of these would strengthen the performance of the trading regime; others would reduce trading incentives. Annex 1 outlines the development of the tradeable quota system and highlights the other important policy measures. Three distinct phases of the manure policy have been identified by Dutch policy makers and researchers, characterised by policy measures to stabilise, reduce and then equalise the nutrient problem.[5]

Institutional setting focussing on government policies

Manure Policy Phase I (1987-90) – measures to stabilise

Under the 1986 Manure Act each farm was required to calculate an annual reference level of manure production in phosphate terms. This was obtained by multiplying the number of animals held on the farm on 31 December 1986 by a set P_2O_5 coefficient for each animal species.[6] The Act also made it illegal for a farm to annually produce more manure than would result in a manure to land ratio greater than 125 kgP_2O_5/ha based on land either owned or in long-term lease. An important exception was provided to existing farms that had a higher ratio on the reference date but these farms were: (a) not permitted to produce more manure than their reference level; and (b) required to pay a tax of EUR 0.11 per kgP_2O_5 above 125 kg/ha and EUR 0.23 per kgP_2O_5 above 200 kg/ha from 1 May 1987.

Farms with a manure to land ratio of less than 125 kgP_2O_5/ha (defined as "manure deficit" farms) or new farms could increase animal numbers until this level is reached. Farms with a ratio above 125 kgP_2O_5/ha ("manure surplus" farms) were much more restricted. They could increase animal numbers by acquiring additional land, but had to acquire enough land to reduce the ratio to 125 kgP_2O_5/ha before they could actually hold more stock. Second, they could increase their manure reference level. However, the manure reference level was not tradeable and was only transferable under certain strict criteria (marriage or heritage or the transfer of the complete farm). Pig and poultry farmers were also given the opportunity to show, through the "mineral input registration system" (MIAR), that the phosphate excreted by their animals was less than the set coefficients by improving their nutrient feeding regime. This incentive was only applicable to reducing the amount subject to levy and did not change their manure reference level.

During the lead-up to the introduction of farm specific manure reference levels LTO Nederland, the Dutch farmer's organisation, protested strongly against their non-tradability. They feared this constraint would limit the expansion, and hence the international competitiveness of livestock farming in the Netherlands (Geluk, 1994). In response the Minister of Agriculture, when introducing the manure production rights, promised to consider the question of tradability. There was considerable debate within the farming community over the issue. Small farmers and those about to commence were against while large farmers were for tradability. Some wanted to exclude certain farms according to their size or location.

Another issue for the government to address was that farmers had anticipated the reference date of 31 December 1986 and had consequently stalled more animals than normal. After 31 December they cut back their herds to normal size. In this way most farmers acquired a higher manure reference level than they actually needed. Officials were concerned that once trading was allowed, farmer would sell these "dormant" rights leading to both a windfall financial gain for them and to an increase in nutrient leakages as the "dormant" rights are used by other farmers.

Manure Policy Phase II (1991-94) – measures to reduce

On 1 January 1994 changes were introduced to allow trading to occur. To signal the change in policy the manure reference level for each farm was renamed as a "manure production right", still expressed in phosphate terms. Rules were designed to either prevent an increase or encourage a decrease in environmental pressure as a result of trading.

- A farm's manure production right was divided into two parts: a land-based quota (derived by multiplying the hectares farmed [owned or in long-term lease] by 125 kg P_2O_5) and a non-land-based quota (the remainder). Trading was only permitted for non-land-based quota.

- To account for the existence of "dormant" rights, the level of phosphate manure production was reassessed for each farm for the years 1988-90 using the same methodology (*i.e.* animal numbers times set coefficients). This did not change a farm's manure production right but if the original reference level was greater than the highest annual level of 1988-90 the difference was not tradeable.

- For each farm, the non-land-based manure quota was allocated to three specific animal categories (cattle and turkeys, pigs and chickens, and others) reflecting the farm specific situation. Trading was permitted within but restricted between animal categories, *e.g.* manure production rights allocated to cattle and turkey could not be purchased for pig and poultry production but manure production rights allocated to pig and poultry could be purchased for cattle production.

- Trading was also geographically restricted. Two surplus regions, with average manure production greater than 125 kgP_2O_5/ha were identified: one in the east (comprising three-quarters of the province of Gelderland, one-third of Utrecht and a large part of Overijssel) and the other in the south (large parts of the provinces of Noord-Brabant and Limburg). Trading in manure production rights could take place within and between these two surplus regions and could be sold to manure deficit regions but rights were not allowed to be brought into these regions from a manure deficit region.

- To reduce the overall volume of manure, 25% of the quota involved in each trade was retired.

- A farmer who purchased additional manure production rights had to certify that they had land (either owned or under two year contract) to apply their total amount of manure at the appropriate land application rates, *i.e.* farms with phosphate manure production greater than 125 kgP_2O_5/ha could expand by purchasing additional rights, but were restricted by the need to comply with land application rates set down by regulations.

The trading rules also stated that quotas could not be leased; users had to be owners. Further, the rules originally stated that the quota system would be terminated on 1 January 1997.[7] This reflected the understanding reached in 1993 between the Ministries of Agriculture and the Environment and LTO Nederland that by 1998 the quota system would be obsolete with the introduction of a farm level nutrient accounting system (Wossink and Gardebroek, 2003).

Manure Policy Phase III (1995 -) – measures to equalise

It became increasingly clear that simply avoiding further increases in the volume of manure was insufficient. The government wanted to create a national balance between the production and disposal possibilities for animal manure. With lower and lower land application rates reducing the quantity of manure that can be spread on land (to try and meet the requirements of the EU Nitrates Directive) and the proven limitations of non-land based disposal possibilities such as processing into biogas and fertiliser or export, a reduction in the overall volume of manure was inevitably required (Henkens and van Keulen, 2001). The government also wanted to lower the volume of manure to reduce the incentive for fraudulent behaviour under the MINAS system, whereby levies would be imposed on surplus levels of both phosphate and nitrogen. Since the manure surplus (calculated at that point to be 14 000 tonnes P_2O_5 and growing as the land application limits fell) was considered to largely originate in the pig sector, measures were specifically introduced to target this.

In 1995, the government announced a 30% reduction in the non-land-based quota for pig and poultry farmers. This was in response to the improved feeding regimes that had been introduced by pig farmers which had lowered the amount of phosphate excreted per animal by 20-25%.[8] This was enacted on 1 January 1996. However, when MINAS was introduced on 1 January 1998, the quota system was not abolished as previously committed and the 30% reduction was revoked.

On 1 September 1998, the *Pig Farming Restructuring Act* came into force, replacing manure production rights for pigs by a system of animal production rights. Separate quotas for fattening pigs and sows were introduced based on the actual number of these animals on each farm in 1995 or 1996. With the aim of completely removing the surplus, the government attempted to initiate a number of measures including an across the board 25% reduction in pig animal production rights, the government purchase of pig animal production rights and an increase in the quantity of animal production rights withdrawn when a farmer quits the sector (from 25% to 40% rising to 60%).

However, pig farmers, who had already expressed their hostility to these proposals during their development, took the government to court and a lengthy litigation battle ensued during which time the measures could not be applied. In January 2000, the court ruled that the government could introduce animal based rights for pigs, implement a generic 10% reduction in those rights (but not the 15% reduction scheduled for 2000), and instigate a buy-out scheme in 2000 and 2001. Manure production rights were also converted to animal based rights for the poultry sector in 2001, with separate rights for laying hens and birds kept for meat production.

The purpose of the buy-out scheme [Regeling Beëindiging Veehouderijtakken (RBV)] was to reduce the manure surplus by giving pig, poultry and fattening calve farmers an opportunity to end their activities. Under the programme the government (with financial support from regional government) offered farmers the replacement value for their livestock, a lump sum payment and a contribution towards the costs of breaking down the stables in an environmentally responsible manner. In total, EUR 250 million was spent during the two years that the scheme was available. Through the purchase of animal production rights (some of which were dormant), 11 890 tonnes P_2O_5 (25 550 tonnes N) was withdrawn, equivalent to 55% of the estimated manure surplus in 2003 (Vliet and Ogink, 2004).

Organisational arrangements

Trading in manure/animal rights takes place in the private sector. Brokers are normally used to establish transactions rather than direct farmer to farmer deals. However, the government, through the General Inspection Service (AID) of the Ministry of Agriculture, Nature and Food Quality (LNV), is involved in approving each transaction to ensure that farmers purchasing the manure/animal rights have an appropriate manure disposal plan. AID, together with the police, also has responsibility for enforcing the manure application rules. AID inspectors (around 150) undertake random sampling, inspection of risk groups and investigation of violations reported by third parties. Farms are likely to be inspected by AID once every ten years (Verschuur *et al.* 2003). In addition to AID, the Agricultural Levies Office, another agency of the LNV, carries out the administrative audits for MINAS and the manure transfer contracts. It verifies MINAS returns and imposes appropriate levies on farmers exceeding the levy-free loss standards.

Measures to promote transactions

One of the specific measures introduced to promote transactions was a brochure explaining the system to farmers, although its length (66 pages) also signalled the complexity of the trading arrangements. Wossink (2000) estimates that the conventional transaction costs incurred by pig producers in trading can be as high as 17% of the average quota price. Efforts to more evenly match quota levels with actual use (such as the non-tradability of dormant rights) were also important in trying to ensure that the market was not oversupplied, but suggest that most farmers had an initial over-allocation, limiting market demand. Further, the fact that emission levels were determined administratively (*i.e.* by multiplying animal numbers by set coefficients) and not by actual emission rates meant that quota supply could not be generated by improving management and technological factors.[9] Consequently farmers willing to sell quota were those reducing animal numbers. In effect the price paid for quota was a transfer from farmers wishing to expand animal production to those looking to reduce animal production. Tradeability provided a windfall financial gain for those looking to leave the sector.

Studies indicate that policy uncertainty regarding the future of the quota system and the introduction of further constraints had a negative effect on the operation of the market (Wossink, 2000; and Wossink and Gardebroek, 2003). Uncertainty surrounding policy changes, such as the permanence of the 30% cut in quota announced in 1995, drastically increased transaction costs, reducing the incentives to purchase quota and thereby reducing the efficiency of the market. The process of administrative approval also slowed the development of the market. In 1994, 37% of the trade application had to be resubmitted because of shortages in the manure disposal plan submitted for approval by the quota purchaser.

It is also important to consider the potential impact of other policy measures on trading. With an increase in the stringency of regulations concerning the land application and storage of manure, demand for manure/animal production rights would decrease (as the cost of expansion increases) and supply would increase (as the increased manure management costs force some farmers to quit), leading to a reduction in the quota price. In addition to manure/animal production rights, livestock farmers wanting to expand production in the two surplus regions must acquire ammonia rights. Trade in ammonia rights is only allowed within a county and so is even more spatially restricted than trading in phosphate quota (Vukina and Wossink, 2000).

Assessment and conclusions

By imposing a farm specific "cap" on the quantity of phosphate manure production, or animal numbers in the case of pig and poultry, the Dutch quota system potentially prevents further increases in environmental pressure. Studies suggest that without the cap pig and poultry animals numbers, and manure production would have been around 10% higher (Wossink, 2004). The cap also provides a stable situation in which other environmental policies can then more effectively operate. Reductions in the national or regional environmental impact can be obtained, for example, through government buy-outs. Regulations relating to the land application of manure can be implemented to reduce the farm-level impact. On the downside, the imposition of a farm specific quota, without the incentive to innovate, causes a freeze in the agricultural structure, hampering adaptation and investment that could assist in solving the environmental problem as well as reducing the long-term economic viability of the sector.

Analysis of the distribution of livestock across regions and sectors in the Netherlands indicates that there has not been any change resulting from the introduction of trading in manure/animal production rights (Annex 2). It seems that while such movements may have been economical viable, trading has been limited by other policy measures such as milk quotas, or local community resistance. In terms of on-farm structural impacts, the trading rules have permitted livestock production to expand in scale without increasing the intensity of production. Farms above the average have reduced their intensity and those below have increased it.

A number of conclusions that can be drawn from this analysis of the Dutch manure/animal production rights system. It is crucial to establish an appropriate base level of quota/rights. The initial over-allocation limited demand, and required additional rules and uncertainties to be introduced. Policy uncertainty relating to trading rules, the termination of the system, etc., then had a negative impact on the operation of the trading system. Consideration must be given to the incentive effects of other policies on trading. While the quota system may have allowed other environmental policies to be introduced, these have generally been to the detriment of the manure trading system. Agricultural policies have also limited the ability of movement in manure production rights to occur as well as non-market factors, such as local community response. Finally, the failure to base rights on actual levels of emissions limited the contribution of the tradeable rights system to enhancing environmental efficiency.

Notes

1. Action was also required as a result of international environmental commitments made by the Netherlands, in particular the 1988 Paris Convention to reduce nutrient supply to the North Sea to 50% of 1985 levels by 1995.

2. One kg phosphate (P2O5) = 1/2.29 kg phosphorus (P).

3. P is less reactive than N, it does not exist in different oxidation states, and is rather insoluble in its dominant natural form in crops, animals, manure, soils and water.

4. Further, the Nitrates Directive required all EU countries to introduce by December 1999 nitrogen specific application rates for manure in areas determined to be Nitrate Vulnerable Zones (NVZ). The Netherlands has declared the whole country as a NVZ.

5. The phasing in over a long period of regulations and stricter requirements etc was done to minimise the socio-economic consequences for the livestock sector. It was originally hoped that alternatives such an manure processing, more efficient feeding and the export of manure would provide sufficient alternative possibilities for solving the problem. However, technologies were not able to solve the problem, forcing policy makers to use on-farm management (*i.e.* MINAS) to solve the problem.

6. Initially applicable to cattle, pigs and poultry. In 1992 it was extended to manure from other animals. Greater details of the technical aspects of the manure production rights system are outlined in Annex 1. This section will discuss more the political and institutional issues surrounding and influencing the development of the tradeable system.

7. Article 22-3 of *Verplaatsingsbesluit* [the "Law regulating transfer of manure production rights"], Staatsblad 171, Staatsuitgeverij, the Hague.

8. There had not been a comparable innovation in poultry although only 80-90% of quota was actually being used on poultry farms which meant that the cut did not have a big impact on them (Wossink, 2004).

9. This limited the ability of the Dutch manure production rights system to contribute to what the OECD terms "soft effects" of tradeable permit systems – incentives to improve firm resource management (OECD, 2004). There is significant evidence that the overall policy mix of regulations, taxes and education contributed to a reduction in emissions, particularly on pig farms through improved nutritional measures. However, these reductions cannot be attributed to the quota system (Wossink, 2004).

Annex 1. Development of Manure Policy Measures

Manure Policy Phase I (1987-90)
Objective: to stabilise manure production and to introduce a national soil protection regime for all activities that threaten to pollute or damage the soil or groundwater

Tradeable manure quota measures	Other measures
Establishment of the manure quota	*Regulations on manure application*
The *Manure Act 1986* introduced a system under which each farm was "grandfathered" a manure reference amount, expressed in kg phosphate (P_2O_5) per year. This was determined by the multiplying the number of animals by a standard coefficient of manure production per animal species. These animal specific standards were calculated as the difference between phosphate supply (in feeds, animals, fertiliser, etc.) and phosphate removed (in meat, milk, eggs, etc.). The difference was assumed to represent the P_2O_5 content in manure of the specific animal category.[1] For cattle, pigs and poultry the assessment took place on 31 December 1986; for sheep, goats, rabbits, ducks, foxes and nutria, 31 December 1991.	The *Soil Protection Act 1986* allowed the establishment of a *Decree on the use of Animal Manure* to place restrictions on the quantity, timing and method of application of manure. The decree has been adapted continuously in the course of time. In Phase I period the main restrictions included the following.
On both occasions assessments were also made of the available land (whether owned or leased for a period of six years or more) for agricultural production. Multiplying this area by 125 kgP_2O_5 gave the farm's assessed acerage-based phosphate rights. Farms whose reference level was less than this were classified as "manure-deficit" farms. These farms were able to expand livestock production to a point were manure production reached 125 kgP_2O_5/ha. Farms whose reference level exceeded this were classified as "manure-surplus" farms. These farms could only expand livestock production by acquiring land until the assessed rights became larger than the reference, *i.e.* it became possible to spread manure at a rate of 125 kgP_2O_5/ha or lower.[2] New farms were allowed a total manure production from all animal sources of up to 125 kgP_2O_5/ha.	To reduce water pollution: (a) maximum applicable rates in kg P_2O_5 per ha/year of 250, 350 and 125 for grass, maize or other arable crops respectively; and (b) prohibitions on spreading manure: on frozen soil and when rainfall is high; on grassland between 1 October and 30 November and during snow cover, and on sandy soils between harvest and 1 November. To reduce ammonia emissions, from 1987 all manure spread on grassland and maize fields must be ploughed into the soil within 24 hours of spreading.
	Levies
	Under the *Manure Act*, farms that produce more than 125 kg/P_2O_5/ha/year are required to pay a levy (a tax of EUR 0.11 per kg P_2O_5 above 125kg/ha and a tax of EUR 0.23 per kg P_2O_5 above 200kg/ha) from 1 May 1987. This money was used to finance the operation of the manure banks and research into ways to solve the manure problem.
Manure book-keeping was required to ensure that actual phosphate production did not exceed the reference amount, and to also ensure that the manure application rates were not exceeded.	*Payments*
Trading of quota was not permitted and reallocation was severely restricted to the following three situations: (a) transfer of whole farm ownership (same place); (b) marriage and inheritance; and (c) with annulment of a lease.	Financial support was provided to farmers wishing to apply innovative techniques to reduce the production of manure or improve the reprocessing or disposal of livestock wastes. Subsidies were also provided for investment in storage capacity, both on individual farms and join storage facilities.
	Extension and advisory service
	Extension services were provided in areas such as nutrient management and reducing the level of water used in production.

1. For example, the phosphate standard for finishing pigs is 7.4, *i.e.* one finishing pig will produce 7.4 kg of P_2O_5 per year.

2. For example, consider a 4 hectare farm with 1 000 finishing pigs on 31 December 1986. The manure reference amount would be calculated as 7 400 kgP_2O_5, with acerage-based rights of 500 kgP_2O_5 (4 x 125). This farm could only begin to expand production by acquiring (through purchase or long-term lease) an additional land 56 hectares of land (60 x 125 = 7 500).

Manure Policy Phase II (1991-94)

Objective: to reduce the mineral losses from agriculture to the environment

Tradeable manure quota measures	Other measures
Trading rules developed for the manure quota	*Tightening of regulations*
On 1 January 1994, changes were introduced to allow trading to take place.	During this phase, existing regulations in place to deal with were tightened.
(1) For each farm, the reference levels were converted into *manure production rights* still expressed in P_2O_5.	To reduce water pollution: (a) maximum admissible application rates of manure were lowered for each of the three categories of land; and (b) the period during which manure is not to be spread on the land was extended to 1 October – 31 January on grassland and arable land on sandy and loess soils and in valleys.
(2) This was divided into two parts: a land-based part (125 P_2O_5 x ha) and a non-land based part (the difference between the manure quota and the land-based part).	
(3) It was also allocated to three classes of livestock in descending order: (i) cattle and turkeys; (ii) sheep, goats, foxes, nutria and ducks; and (iii) pigs and chickens.	To reduce ammonia emissions: (a) as of 1 January 1991 the injection manure into the soil was obligatory on grassland in some sandy parts of the country. The obligation to inject manure was extended to all grassland on sandy soils in 1992 and to all grassland in 1994; and (b) from 1991, it was obligatory to roof over manure storage silos constructed after 1 June 1987, or to take measures equivalent to roofing.
Only the non-land based quota could be allocated and became tradable with certain restrictions (e.g. certain restrictions on trade across animal categories and between regions, and with the government taking 25% of the quota involved in each transaction).	

Manure Policy Phase III (1995-)
Objective: **to equalise the input and output of minerals in agriculture by reducing nitrogen (N) and phosphate (P_2O_5) losses to an environmentally acceptable level**

Tradeable manure quota measures	Other measures
Refinements to the manure quota For pig and poultry producers the system of farm manure production rights (expressed in P_2O_5) was replaced by a system of *animal production rights* in 1998 and 2000 respectively. For pigs, this was based on the average number of animals in 1995 or 1996, and for poultry producers, the average for the period 1995-97. Within the overall pig quota for a farm, a separate maximum was set for breeding sows based on the number of breeding sows in 1995 or 1996. The same trading restrictions that apply to farm manure production rights also apply to the animal production rights. Farmers in the south and eastern regions where pig and poultry farming is concentrated cannot buy quota from other regions and farmers in other regions can only buy rights if strict environmental conditions are met. A number of steps were also taken to reduce the volume of manure. First, pig production rights were reduced by 10% for each farm. Second, if a farmer ends production, the government took an increased share of the animals (40% to 60%) to reduce the national herd, with the farmer able to sell the remainder to other farmers. If they cannot find a buyer the government will buy the quotas at the market value. In 2000 and 2001, the government offered to buy-back any quota if the farmer wished to quit farming.	*Introduction of MINAS* To regulate the use of nutrients, farmers must enter into Minneral Accounting System (MINAS), which requires farmers to submit annual minerals accounts (balances) showing input and output of both nitrogen and phosphate at the farmgate level. MINAS was introduced in three stages. For 1998 and 1999, it was only compulsory for all livestock farms with more than 2.5 LU per ha (about three-quarters of dairy farms and nearly all pig and poultry farms). For 2000 it was compulsory for all livestock farms. And from 2001 all farms (including arable and horticultural farms) had to participate. For MINAS, the nutrient content of manure is determined by laboratory analysis. Levies are imposed on N and P_2O_5 surpluses above a certain level per hectare. Overtime the levy-free surplus levels have been lowered and the levy rates increased. From 2003, the phosphate levy free surplus level (loss standard) is 20 kg P_2O_5 /ha/year. For nitrogen, the loss standard is 100 kgN/ha/year on arable land (60 on peat soils), and 180 kgN/ha/year on grassland (140 on peat soils). Levies on surpluses exceeding these less standards are EUR 9 kgP_2O_5 and EUR 9 kgN. *Manure transfer contracts* With the introduction of MINAS the Dutch government removed requirements setting maximum allowable application rates for manure. However, in order to try and comply with the requirements of the Nitrates Directive, the government introduced manure transfer contracts. Under this system farmers must ensure that they have sufficient land on which they can potentially dispose of the manure at the application rates of 170 kgN on grassland and 210 kgN on arable land.[1] If a farmer has more manure than can be applied on their land they must arrange a contract with other farmers to dispose of the excess manure. It was introduced in phases, and has been applicable to all farms since 1 January 2002. The calculation is based on the number of animals and a statutory fixed rate of nitrogen production per animal species. *Education/extention* Demonstration and information programmes were introduced to support the introduction of MINAS. *Regulations* New requirements to reduce ammonia emissions, particularly from housing facilities were introduced in 2002.

1. The 250 kgN/ha application rate for grassland was still 80 kgN/ha above the limit set by the Nitrates Directive. The Netherlands sought a derogation for this, arguing that the long growing seasons with high nitrogen uptake and the overall focus of MINAS on nutrients from all sources would ensure safe drinking water standards at met. In October 2003, the European Court of Justice ruled that the MINAS policy failed to meet the requirements of the Nitrates Directive (91/676/EEC). As a consequence, a simpler system based on strict limits on the maximum application of manure nitrogen per hectare will replace MINAS in 2006.

Annex 2.

Analysis of Tradeability

The issue under consideration here is what the provision of tradability can contribute to environmental performance. Two possibilities are considered. Has tradability resulted in an environmentally better geographical and sectoral distribution of manure production in the Netherlands? Has it ensured that expansion of production has occurred without increasing environmental pressure? One of the difficulties in analysing these possible impacts is that changes in production structures and management practices in the Dutch livestock sector have been heavily influenced by the other measures introduced as part of the overall manure policy mix.

In terms of the distribution of manure production, the environmental concern relates to the concentration of pig and poultry production on the sandy soils of eastern and southern parts of the Netherlands. Hence rules were designed to allow only changes in the geographical and sectoral distribution of livestock production that were considered to be environmental beneficial, *e.g.* allowing the rights to shift out of but not into the manure surplus regions, and out of but not into pig and poultry production. In the Netherlands, the standard livestock unit equates to the phosphate loading of one dairy cow, so changes in cattle units by region and sector are a good representation for changes in phosphate manure production.

Over the period 1992-2001 there has been a 13% reduction in the total number of livestock units in the Netherlands (Figure A2.1). However, while livestock numbers have fallen on dairy and other cattle farms, they have increased on pig and poultry farms. Consequently, the share of total livestock on pig and poultry farms has risen from 21% and 11% respectively in 1991 to 25% and 13% in 2001. The share of livestock on dairy farms, the largest livestock farming sector in the Netherlands, has fallen from 43% to 40%. Clearly, there has not been a shift in livestock units away from what is considered to be the most intensive forms of production although during the last few years (2002-2004) the number of pigs and poultry has fallen, from 13 million pigs in 2001 to 11 million in 2004, and from 103 million to 84 million poultry.

Figure A2.11. Livestock numbers by type of farm, 1992-2001

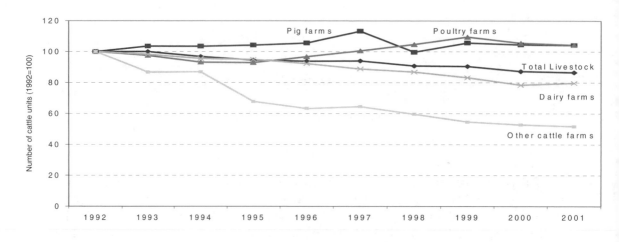

Source: Central Bureau of Statistics, www.cbs.nl, data downloaded August 2004.

One of the major constraints to this shift occurring is the existence of milk quotas, which limits the expansion of the dairy sector. While dairy production has expanded at the farm level, requiring the purchase of additional manure production rights, and these may have be purchased from pig and poultry farmers, the overall level of milk production, and hence manure production rights required by the dairy sector has not increased. In fact, with a fixed limit on overall milk production, productivity improvements in terms of yields per cow will decrease the number of animals required, and therefore with fixed emission coefficients reduce the volume of manure production rights required.

There has not been a movement towards a more environmentally favourable geographic distribution of livestock production (Figure A2.2). The rate of decrease in livestock numbers has been similar in the two manure surplus regions as in the rest of the Netherlands combined, although some of the other provinces have seen an increase in livestock numbers. Consequently the share of livestock in the manure surplus regions has remained at around 56%.

Figure A2.2. Regional distribution of livestock
1991-92 and 2000-01

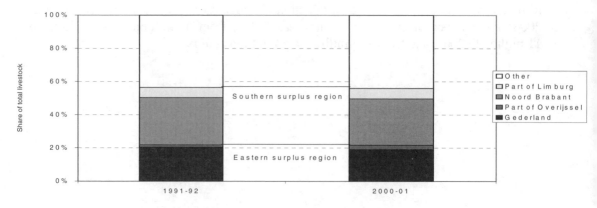

Source: Central Bureau of Statistics, www.cbs.nl, data downloaded August 2004.

In their investigation of the trading regime using data from 1986-96, Vukina and Wossink (2000) estimated that pig farmers in the surplus regions have an economic incentive to sell their quota and buy land in the deficit regions because of the higher quota price and cost of environmental compliance in the surplus regions. They concluded that over time the movement of pig farmers would lead to an equalisation of production costs across regions, elimination of rents and the convergence of land prices. While some farmers have made this migration, it does not appear to have happened to any large degree. One of the major factors that appear to have limited this movement is the strong and active resistance of local communities in manure deficit regions to the establishment of intensive pig and poultry farming operations (Mass and Wisserhof, 2000).

In order to maintain competitiveness, Dutch livestock producers are under continuous pressure to increase the scale and intensity of their production. The tradeable rights system requires those that expand to do so in a more environmentally friendly manner. Analysis of changes in the scale and intensity of pig production indicates that tradability has resulted in an expansion of the sector in a manner that is likely to be placing less pressure on the environment and illustrates how they have achieved this.

As mentioned above livestock numbers on Dutch pig farms have increased over the period 1992-2001 by around 4%, from 1.3 to 1.34 million cattle units (Table A2.1). At the same time there has been a 32% decrease in the overall number of pig holdings, leading to a 50% increase in the scale of production as measured by the number of animals per holding. More importantly from an environmental point of view there has been only a slight increase in the total area of these operations and the overall level of intensity of production on all types of pig farms, as measured by the number of livestock per hectare. At the sectoral level pig farmers have been able to increase the number of livestock they are carrying on roughly the same area of land in a manner that has limited the environmental impact. How have they achieved this?

Table A2.1. Developments in the scale and intensity of Dutch pig farms

Type of pig farm	Year	Number			Scale		Intensity
		Animals (cattle units)	Holdings	Area (ha)	Animals/holding (cattle units)	Area/farm (ha)	Animals/ha
Total of 3 types	1992-93	1 303 886	8 471	38 607	154	4.56	33.8
	2000-01	1 336 751	5 772	38 846	232	6.73	34.4
Breeding	1992-93	439 805	3 187	15 423	138	4.84	28.5
	2000-01	433 528	1 964	13 416	221	6.83	32.3
Fattening	1992-93	396 806	3 328	8 547	119	2.57	46.4
	2000-01	380 104	2 233	9 628	170	4.31	39.5
Other	1992-93	467 274	1 956	14 637	239	7.48	31.9
	2000-01	523 118	1 576	15 801	332	10.03	33.1

Source: Central Bureau of Statistics, www.cbs.nl.

A closer look at changes in the number of animals and area by type of pig farm provides the answer. Simply, pig farmers have increased intensity on farms with a lower than average intensity of production and decreased it on those with a higher than average intensity. In terms of the denominator there has been a reduction in the area of land on breeding farms (where intensity is lower) and an expansion on fattening farms (where intensity is higher). In terms of the numerator, there has been a greater reduction in the number of animals on fattening farms (higher) than on breeding farms (lower). Furthermore, there has been an increased in the number of animals kept for slaughter (fattening) on breeding farms. Both the number of animals and area has increase on "other" pig farms. As a consequence the intensity of production on the three types of farms is converging towards the average (Figure A2.3). By increasing intensity on farms below the average and decreasing it on farms above the average, Dutch pig farms have been able to expand the number of livestock while maintaining a relatively constant area of land in, and intensity of, production.

Figure A2.3. Intensity of production on Dutch pig farms
1992-2001

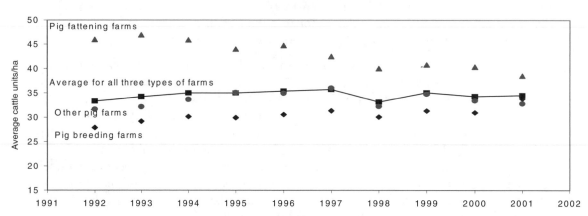

Source: Central Bureau of Statistics, www.cbs.nl, data downloaded August 2004.

14. Wetland Mitigation Banking (Type N1)

Abstract and classification

Wetland mitigation banking aims to compensate for wetland loss due to land development. Developers purchase mitigation credits from a mitigation bank, which is established by a third party that creates mitigation credits through the restoration/construction/enhancement of wetlands and sells credits on the market. The agricultural sector is a main actor in this area in that it converts its own wetlands to other land uses. Since the exchange of transferable credits on the market is the core mechanism used to increase economic efficiency, this case study is categorised as Type N1.

Background

The agricultural sector has had a significant impact on the loss or gain of wetlands. By the mid-1980s, the United States had lost half of all its wetlands, and the agriculture sector was responsible for up to 54% of those losses in the decade starting from the mid-1970's (Dennison, 1997). In the last 25 years, various policy instruments have sought to discourage wetland conversion by withdrawing conversion incentives, regulating conversion through water quality and other legislation, and funding voluntary programs to restore wetlands (Heimlich *et al.* 1998). As a result, between 1997 and 2002 there was a significant gain in wetland area originally lost due to agriculture (Natural Resources Conservation Service, 2004). It is difficult, however, to separate policy and market factors responsible for this wetland gain since low commodity prices have also discouraged conversion.

Section 404 of the Federal Water Pollution Control Act of 1972 (known as the Clean Water Act) provides legislation that limits the loss of natural wetlands and allows developers who remove wetlands in one region, to mitigate these losses by conserving wetlands in another region. The Act regulates discharge of dredge and fill materials, but does not specifically regulate wetland drainage or clearing. Therefore, all wetland conversion activities cannot be covered by the Act.

The idea of mitigation banking originated in the 1970s at the same time as the Act concerning wetland development regulations. Until the 1990's, the US Army Corps Engineers (Corps), was primarily responsible for the implementation of the Act and it issued wetland development permits to developers who took off-site compensatory mitigation measures. Meanwhile, the Environmental Protection Agency (EPA), was authorised under the Act to object to the issuance of permits. Its priority was to avoid any wetland loss and minimise the negative impacts on existing wetlands, rather than replacing them with newly created wetlands. These conflicting approaches hindered the development of the mitigation banking scheme. It was also eventually found that off-site mitigation measures failed in general because they were poorly designed, used inefficient construction techniques, and lacked proper monitoring and maintenance schedules (Studt and Sokolove, 1996). Taking into account these failures, the Army Corps and the EPA agreed in 1989 to adopt EPA guidelines that prioritised the avoidance and minimisation of impacts on existing wetlands, and off-site mitigation became the last instrument to be considered. This compromise smoothed appropriate mitigation banking applications.

Negative externalities addressed and the mechanism used

Wetlands have various ecological, biological and recreational functions that benefit society. These include wildlife habitat formation, the conveyance and storage of floodwater, the prevention of erosion and saltwater intrusion, sediment control, water supply and quality maintenance, and recreational activities such as fishing and wildlife observation. Agriculture has had negative effects on such wetlands, including through the conversion to farmland by drying or filling. It has drained polluted waters or sediments into wetland areas. As negative as these practices are, there are also cases where activity on farmland can be harmonised with activities on adjacent wetlands by providing wildlife habitats and maintaining hydrological configuration, thus providing a better environment than converting to residential or industrial uses. The farming sector participates in the mitigation banking scheme as a provider of converted farmlands to be restored to healthy wetlands or credit applicants who no longer use wetlands for agricultural purposes.

Developers who cannot avoid destroying wetlands must obtain a development permit from a regulatory authority and are required to implement compensatory mitigation measures as prescribed in the Clean Water Act. Their options are thus their own on-site or off-site mitigation, or the purchase of mitigation credits from a third party. The first approach requires that the developer mobilises his own technical expertise to restore or create wetlands within or outside the project area. Otherwise, a third party, a mitigation banker likely provides the necessary expertise for wetland creation or restoration more efficiently than the developer, and issues its credits which are divided into acres and sold as such to developers. This transaction legally transfers the liability for wetland mitigation from the developers to the bankers.

Although anyone can submit an application to a mitigation bank, it is subject to the regulator's assessment and approval. The application includes a plan defining the responsible bodies and required costs for the development and long-term maintenance of restored/created wetlands. Private firms, environmental NGOs, and regulatory agencies have performed as bankers. Government bodies make and enforce trading rules, and monitor compliance. Key rules include standard design and construction, performance, monitoring and maintenance, long-term management, time to market, and liability for failure. Federal and state regulatory agencies form a Mitigation Banking Review Team (MBRT), which enforces the rules. Once a wetlands project is approved, a limited loss of wetlands is allowed before the wetland construction by the banker. This means a temporary loss of wetland functions associated with the debiting of projected credits.[1] A credit applicant, before applying for the credits, must provide MBRT with justification as to why his development project cannot avoid impacting on wetlands and evidence that the use of bank credits is the most environmentally preferable measure. MBRT monitors the banker's performance on the wetland construction, and enforces any remedial measure on the development project whenever necessary.

Advantages of the mitigation bank scheme include:

- Consolidation of fragmented compensatory mitigation into a single large parcel or contiguous site that strengthens the ecological benefits for wetlands.

- A mitigation bank brings together financial resources and technical expertise, thereby increasing the potential for the successful long-term management of mitigation; and

- The use of mitigation banking by multiple applicants reduces the total permit processing time, and agency resources to review and monitor individual projects.

The basic mechanism for both mitigation banking and tradable permit schemes is that through an exchange of permits (credits), pollution abatement (wetland mitigation) is entrusted to an entity which has a comparative advantage for doing this, which increases the efficient use of resource. The major differences are as follows.

- The total amount of pollution permits allowed is capped for the tradable permit scheme, while there is theoretically no limitation on the issuance of credits as long as there is a demand for wetland development and mitigation. In practice, however, the issuance of credits is constrained by regulations, *e.g.* credit trading should be made within the same watershed. Further, it seems that demand for wetland mitigation is not significant.

- Unlike general tradable permit markets, the credit market is small so that the price is still far from determined in competitive circumstances. In most cases, there is only one credit supplier and a small number of applicants in the credit market.

Wetland mitigation banking pilot project designed for agriculture in Missouri

The financial as well as the ecological feasibility is important in the design and operation of the mitigation banking scheme. Particularly, as the cost of restoration/creation of wetlands is not low, the resulting credit price can be high in order for the banker to collect on capital investment. Land development projects for residential, commercial and public infrastructure land uses are the main mitigation users because they can pay for relatively high priced credits due to high returns from development investment. In contrast, the agricultural sector is not a large credit user.

The first wetland mitigation banking project specifically designed for agriculture was started in 1999 in Missouri. The targeted wetlands were small and scattered with less ecological value, rendering it problematic for farming and reducing its productivity. The mitigation site was 73 acres of previously-converted croplands. The area was restored to a complex of seasonal shallow water areas, restored forest, and scrub-shrub wetlands. The landowners as well as owners of the adjacent areas sought the restoration. The mitigation banker in this case was a private organisation affiliated with American Farmland Trust, a charitable non-profit organisation.

The cost of mitigation is the sum of the easement acquisition,[2] site restoration, short term (3-5 years) mitigation monitoring, long term land management, project administration, performance bond,[3] etc., which were estimated at about USD 167 000. The price per acre of mitigation credit was USD 2 300, which was calculated as USD 167 000 divided by 73 acres. In contrast, the credit prices of other non-agriculture projects were about UDS 50 000 to USD 100 000 per acre (Environmental Law Institute, 2002). Credit applicants bought the permit(s) prior to the commencement of restoration works. Once sufficient funds were in place, the mitigation banker began the work. A mitigation bank review team (MBRT) representing relevant government agencies monitored the process and was authorised to direct the scheme.

The landowner conducted maintenance activities, guidelines for which were agreed with the mitigation banker. A USDA regional office monitored them on an annual basis until the wetland reached an ecologically sound and self-sustaining level. If performance standards could not be achieved, remedial measures such as replanting or hydrological alternation were taken by the landowner.

Institutional setting focusing on governmental policies

The mitigation banking scheme is administered under the Clean Water Act 1972 in conjunction with other federal and state laws, guidelines, and programs. Of these, of great relevance to wetlands and agriculture are the Food Security Act 1985 and federal guidelines for mitigation banking, as well as the Clean Water Act 1972.

The primary federal law that regulates wetland development activities is Sections 401 and 404 of the Clean Water Act, although there are no detailed provisions concerning mitigation policies or compensatory measures. EPA's guidelines pursuant to the law instead define "sequencing" mitigation measures: (i) the potential adverse effects on existing wetlands should be avoided as much as possible, (ii) practical measures to minimise the unavoidable effects should be taken, and (iii) appropriate and practical measures to compensate the adverse effects should be taken. This means that compensatory actions including creation of man-made wetlands are permitted after all avoidance or minimisation measures are exhausted. In addition, the policy goal underlying the concept of a compensatory approach is no net loss of wetlands,[4] which means that wetland losses must not be more than wetlands gains. The areas and quality of restored/created wetlands are important elements to examine in compensatory actions including mitigation banking. Further, ongoing and "normal" agricultural practices in wetlands are exempt from permit requirement under the Clean Water Act.

Agricultural actors also regulate the conversion of wetlands to farming use through the so-called "Swampbuster" provision of the Food Security Act of 1985 and the Food, Agriculture, Conservation and Trade Act of 1990. This provision discourages conversion by denying eligibility for a wide range of farm program benefits on all acres operated by a grower who either converts a wetland or plants on a converted wetland. A farmer violating this measure can regain eligibility for future program benefits by restoring the converted wetland to its original condition. This provision does not prohibit the conversion, but only removes eligibility for some grant benefits. Therefore, this law is less effective where program participant rates are relatively low, and where the number of program participants is variable reflecting market conditions. Heimlich (1988) stated that although the provision might have halted the conversion of about 6 million acres, it was likely to be ineffective on another 6.3 million acres. There were other weaknesses in the provision, such as exclusion of agricultural utility land use in the converted wetlands, but they have been amended accordingly.

There are also other programs and polices in relation to agriculture to protect wetlands. The Wetland Reserve Program authorised by the Agricultural Improvement and Reform (FAIR) Act of 1996, and re-authorised in the Farm Security and Rural Investment Act of 2002, provides financial and technical assistance to landowners who enhance wetlands in exchange for retiring marginal lands from agriculture. The Conservation Reserve Program also contributes to the enhancement of wetland resources by reducing soil erosion and sedimentation.

The federal government, reflecting the growing use of the mitigation banking scheme, issued Federal Guidelines for the Establishment, Use and Operation of Mitigation Banks in 1995. These guidelines are comprehensive in terms of giving detailed guidance on all aspects of mitigation bank exercises, including the procedural framework. One of the notable points is that the guidelines clarify the role of the mitigation bank review team, which includes the US Army Corps and the EPA, and which verifies and monitors the bank's performance. The concurrence of the team is essential for the scheme to proceed. Another point is that due to the sensitivity of wetland ecology, site selection, technical feasibility, long term management and responsibility, and financial assurance are emphasised.

Organisational arrangements

On-site mitigation schemes require developers to mitigate wetland loss, but they pursue the least cost mitigation measures to maximise their profits from land development, which are the cause of an observed poor success rate (Shabman *et al.* 1996). In contrast to on-site mitigation shortcomings, off-site mitigation schemes introduce credit transactions, in which credit suppliers can concentrate their resources to restore wetlands but in a balanced manner between the least costly and the more ecologically sound approaches. As many suppliers are NGOs or private firms conscious of the environment, the pursuit of this balance is not in contradiction with the organisations' objective. Nonetheless, various regulations have to be imposed, such as reporting and long-term maintenance, to secure success in ecological terms; indeed it is possible that credit suppliers could prioritise financial gains. These regulations likely raise the transaction costs of the scheme. The balance between costs and the outcome is vital to the scheme. Shabman *et al.* (1996) stated that credit markets could not exist in the absence of governmental regulations that also create the demand for wetland development permits that in turn require compensatory mitigation credits.

Mitigation bankers are subject to regulatory approval. Ownership ranges from private profit-seeking investors, to private non-profit NGOs, and government agencies, according to project profitability, risk, and public benefits. Recently, no more than 10-20% of the total credits have been provided by private investors because of regulatory conditions that have created barriers to market entry (Shabman, 2004). For the two agricultural cases examined, due to low profitability, a private non-profit organisation affiliated with American Farmland Trust served as a mitigation banker. No profit was made and a fee-based mitigation system was pursued in both cases. Where there is a high expectation of a profit gain and less risk of failure, private capital investment is likely to be mobilised regardless of the amount.

Costs and difficulties encountered in getting the mechanism to work

Wetland mitigation may fail, but the "success" of mitigation, cannot be clearly defined because there is no general basis on which to judge if the complex process of restoring or creating wetlands will lead to a self-sustaining natural environment. Wetlands are dynamic systems whose hydrology and vegetation change slowly and, no two wetlands are the same. This ambiguity is the major source of failure, although human errors and unforeseen events, such as flooding and infestation are also sources of failure.

The initial assessment of the wetlands both restored and filled is a critical process because its outcome is determinant in deciding the continuation of the entire process. For example, the type of wetlands created and the techniques used will determine the probability of success and the trading ratio (see next paragraph) will affect profits earned. However, Shabman *et al.* (1996) stated that "wetland functions are notoriously difficult to measure individually and cumulatively in any qualitative and quantitative way, and there is no one generalised or "correct" assessment methodology that is applicable to all wetland types and landscape settings." Each mitigation project builds a credit valuation protocol and employs the assessment method that is most suitable to the specific conditions of the project. Furthermore, because the more technically sophisticated methods are generally more expensive, permit applicants and mitigation bankers tend to use the simplified methods.

Mitigation credits serve as a unit of exchange for the provision of compensatory mitigation, and its ecological valuation becomes the basis of a trading ratio that determines units of permitted wetland loss. If the replacement is more or less compatible in function and quality of wetlands, its trading ratio is 1:1. In the Missouri case, filled wetlands were scrub-shrub so that its ratio is 1.5-3: 1 (*i.e.* 1.5-3 acres of a new wetland is exchanged with 1 acre of a filled wetland). Trading ratios are crucial to permit applicants as they determine the credit procurement cost. On the other hand, the ratio works as an insurance against the risk of failure because a higher ratio means more wetlands are restored/created. Regulators are responsible for the determination of the ratio, and are motivated in obtaining a higher ratio.

Performance bonds can be employed to ensure the success of wetland restoration or creation. The suppliers lodge a certain amount of money with the regulators and the money is refunded with interest when success is certified. In case of failure, the money is used to repair the wetland at risk. The amount of the bond would reflect the regulator's best estimate of the cost to repair the mitigation site in case of failure. This approach is effective in reducing the probability of failure, but extra costs are shouldered by credit suppliers and credit buyers.

The cost of restoration/creation of wetlands differs significantly depending on the type, location, and condition of the wetlands. For example, bottomlands and mangroves are ecologically important but require higher costs compared to marsh lands. A change in hydrological configuration or the installation of a new water source is costly. Mitigation bankers try to reduce costs in order to maintain financial viability, while not sacrificing the quality of wetlands. Such a balance, however, is difficult to achieve.

Assessment of institutional and other arrangements and conclusions

The mitigation banking scheme develops only if regulatory policy imposes compensatory mitigation and a goal of no net loss. Under these conditions, the scheme is efficient in bringing financial and human resources together and is ecologically effective in replacing small and scattered wetlands with large, contiguous ones. It is similar to a tradable pollution permit scheme but the big difference is that polluters have pollution rights that are capped for the tradable permit scheme whereby pollution and destruction of wetlands must be offset by the creation of replacement wetlands in the case of mitigation banking.

As mentioned above, the applicability of the mitigation wetland scheme to the agricultural sector seems to be limited due to the high cost of credits. However, the idea and experiences in trading credits could be applied to other agricultural pollution schemes. In the US alone, various pollution trade schemes for the agricultural sector have been attempted or implemented, such as for phosphorous, nitrogen, and sodium controls (Sohngen, 1998).

There are many non-profit organisations, such as land trusts and environmental NGOs, whose priorities include wetland protection. The acquisition of a land title or conservation easement is the main strategic instrument to attain their objective, while mitigation banking, sometimes combined with easement arrangements, is a cost-saving measure used to pursue that objective. Wiebe *et al.* (1996) stated that non-profit organisations offer flexibility and agility, the ability to mobilise private financial and political support, and the capacity to provide local knowledge and insight. Their entry into the wetland mitigation market could expand the market size and bring wider expertise to this area. Mitigation banking can be embraced as a market based solution, although it is clear at this early stage in the mitigation process that trading rules set by regulators are necessary. This could lead, however, to high transaction costs and hinder new players from entering the market.

Notes

1. Credit is a unit of measure representing the accrual or attainment of aquatic functions at a mitigation bank; the measure of function is typically indexed to the number of wetland acres restored, created, enhanced or preserved. Debit is a unit of measure representing the loss of aquatic functions at an impact project site (Federal Guidance for the Establishment, Use and Operation of Mitigation Banks).

2. The landowner sold the conservation easement of his/her property to the mitigation banker, of which price, USD 1 500 per acre to permanently release the development right, ensuring the sustainability of wetland.

3. Performance bonds are explained later in this case study.

4. This was adopted as a policy goal of both the Bush and Clinton administration's wetland plan in 1991 and 1993.

References

Agriturist (2003), "L'agriturismo tiene e riflette", *Notizie da Agriturist*, 24 November.

American Farmland Trust website, www.farmland.org, accessed July 2004.

American Farmland Trust and Agricultural Issues Center (2003), *"A National View of Agricultural Easement Programs"*.

Atzeni, G.O. (2004), *"L'atttività del gruppo di Azione Locale "Capo di S. Maria di Leuca" nel settore dell'agriturismo"*.

Aurland Naturverkstad (2001), *Verdiskapende geitehald i Flåmsdalen. Rapport frå landskapspleietiltak sesongen 2000* (in Norwegian). Annual report 2000 for the landscape management project in the Flåm Valley. Aurland Naturverkstad BA, Report 06.03.01, pp. 8.

Aurland Naturverkstad (2002), *Verdiskapende geitehald i Flåmsdalen. Rapport frå prosjekt for ammegeit og landskapspleie sesongen 2001* (in Norwegian). Annual report 2001 for the landscape management project in the Flåm Valley. Aurland Naturverkstad BA, Report no. 1/02, pp. 12.

Aurland Naturverkstad (2003), *Verdiskapende geitehald i Flåmsdalen. Rapport frå prosjekt for ammegeit og landskapspleie sesongen 2002* (in Norwegian). Annual report 2002 for the landscape management project in the Flåm Valley. Aurland Naturverkstad BA, Report no. 10/03, pp. 12.

Australia, Department of Agriculture, Fisheries and Forestry of Australian Government's website (2004), www.affa.gov.au accessed August 2004.

Barbier, M. (1997), Quand le pollué et les pollueurs se découvrent conventionnalistes, *Revue Française de Gestion*, 112 , pp. 100-107.

Brossier, J. and M. Gafsi (1997), Farm Management and Protection of Natural Resources: Analysis of Adaptation Process and Dependence Relationships, *Agricultural Systems*, 55(1), pp. 71-97.

Brossier, J. and M. Gafsi (2000), Une gestion négociée d'un problème de pollution : pratiques agricoles et qualité de l'eau, l'exemple de Vittel, *Comptes rendus de l'Académie d'Agriculture de France*, 86(2), pp. 57-72.

Centraal Bureau voor de Statistiek (CBS) (2004), *Monitor Mineralen en Mestwtgeving 2004*, Voorburg/Heelen, Nethelands, www.cbs.nl/nl/publicaties/publicaties/bedrijfsleven/landbouw-visserij/j-64-2004.pdf, (in Dutch).

Chia, E. and N. Raulet (1994), Agriculture et qualité de l'eau : négociation et rôle de la recherche - Le cas du programme AGRE, *Etudes et recherches Systèmes Agraires et Développement*, 1994, N°28, pp. 177-193

Chia, E., J. Brossier, and M. Benoît (1992), Recherche-action: qualité de l'eau et changements des pratiques agricoles, *Economie Rurale*, N°206:, pp. 29-30.

Chia, E. (2004), Personal communication.

Coase, R.H. (1960), "The Problem of Social Cost," *Journal of Law and Economics*, Vol. 3, pp. 1-43.

Dalziel, Paul and Lattimore Ralph (2001), *The New Zealand Macroeconomy: A Briefing on the Reforms and Their Legacy*, Melbourne: Oxford University Press.

Daniels, T. (2000), "Saving Agricultural Land with Conservation Easements in Lancaster County, Pennsylvania" in J.A. Gustanski and R.H. Squires (eds.), *Protecting the Land: Conservation Easements Past, Present, and Future*, Island Press, Washington D.C.

Deffontaines, J.P., M. Benoît, J. Brossier, E. Chia, F. Gras, M. Roux (1993), *Agriculture et qualité des eaux souterraines (AGREV), Diagnostic et propositions sur le site Vittel*, INRA-SAD, p. 355.

Deffontaines, J.P., and J. Brossier (2000), Système agraire et qualité de l'eau. Efficacité d'un concept et construction négociée d'une recherche, *Nature Science Société*, N°8(1), pp. 14-25.

Dennison, M.S. (1997), *Mitigation Banking: Mitigation banking and Other Strategies for Development and Compliance*, Government Institutes, Rockville, Maryland, US.

Diehl, J. and Barrett, T.S. (1988), *The Conservation Easement Handbook,* the Trust for Public Land and the Land Trust Alliance, San Francisco, California.

Ducks Unlimited (2004), *Conservation Easement: Protection in Perpetuity*, downloaded from http://www.ducks.org/

Dwyer, Janet and Ian Hodge (1996), *Countryside in Trust, Land Management by Conservation, Recreation and Amenity Organisation*, John Wiley and Sons, UK.

Elliot, R. and R.D. Anderson (1995), *Living with the high country: looking past the end of the Rabbit and Land Management Program*, New Zealand Ministry of Agriculture and Fisheries, Wellington, New Zealand.

Environmental Law Institute website, www.eli.org, accessed June 2004.

European Commission (1990), *Community Action to Promote Rural Tourism —Communication from the Commission* (COM/90/438).

Flåm Development (2004), Presentation of Flåm Development Ltd. and Fretheim Hotel (In Norwegians). Flåm Development Ltd.

Gafsi, M. (1999), "Aider les agriculteurs à modifier leurs pratiques – Eléments pour une ingénierie du changement", INRA SAD, *Façsade*, N°3 , pp. 1-4.

Gardebroek, C. (2001), "The impact of manure production rights on capital investment in the Dutch pig sector", paper presented to the American Agricultural Economics Association meeting, Chicago, Illinois, http://agecon.lib.umn.edu/cgi-bin/pdf_view.pl?paperid=2510&ftype=.pdf.

Geluk, A. (1994), *Economic instruments and new policies in the Netherlands*, Ministry of Agriculture, Nature Management and Fisheries, the Hague (paper prepared for the OECD Experts Meeting on Economic Incentives for Achieving Environmental Goals in Agriculture, Paris, 7-8 June 1994).

Gustanski, J.A. (2000), "Protecting the Land: Conservation Easements, Voluntary Actions, and Private Lands", in J.A. Gustanski and R.H. Squires (eds.), *Protecting the Land: Conservation Easements Past, Present, and Future*, Island Press, Washington D.C.

Hackl, F. and G.J. Pruckner (1997), "Towards More Efficient Compensation Programmes for Tourists' Benefits from Agriculture in Europe", *Environmental and Resource Economics*, 10/1997.

Harvey, David (2004), *How does economics fit the social world*, Presidential address to the 79th annual Agricultural Economics Society Conference on 4-6 April 2004, www.aes.ac.uk/downloads/conf_papers_04/Harvey.pdf, accessed April 2004.

Heimlich, R.E. (1988), *The Swampbuster Provision: Implementation and Impact*, proceedings of the National Symposium on Protection of Wetland from Agricultural Impacts, U.S. Fish and Wildlife Service, Department of Interior, Washington D.C., US.

Heimlich, R.E. *et al.* (1998), *Wetlands and Agriculture: Private Interests and Public Benefits*, Economic Research Services/USDA, Washington D.C., US.

Henkens, P. and H. van Keulen (2001), "Mineral policy in the Netherlands and nitrate policy within the European Community", *Netherlands Journal of Agricultural Science*, Vol. 40, pp. 117-134.

Holland, M. and Boston, J (eds.), (1990), *"The Fourth Labour Government: Politics and Policy in New Zealand "*, Auckland, Oxford University Press.

Hoop, D. de, F. Hubeek and J. Van der Schans (2004), *Evaluatie van Mestafzetovereenkomsten en Dierrechten: Studie in het kader van evaluatie meststoffenwet 2004*, The Agricultural Economics Research Institute (LEI), Report 3.04.03, The Hague, www.lei.nl, (in Dutch).

Hopkins, J. and R. Johansson (2004), "Beyond Environmental Compliance: Stewardship as Good Business*"*, *Amber Waves*, Volume 2, Issue 2, Economic Research Service of the US, www.ers.usda.gov/amberwaves/ accessed September 2004.

Hovorka, G. (2002), Die EU Ausgleichszulage für benachteiligte Gebiete [EU Compensation payments for less-favoured areas], *Facts and Features*, Bundesamt für Bergbauernfragen, Vienna.

Hubeek, F. and D. de Hoop (2004), *Mineralenmanagement in beleid en praktijk: en evaluatie van beleidinstrumenten in de meststoffenwet (EMW 2004)*, The Agricultural Economics Research Institute (LEI), Report 3.04.09, The Hague, www.lei.nl, (in Dutch).

Hubert, B. and J. Brossier (2001), Integration of Bio-technical, Economic and Social Science, *Cahiers d'études et de recherches francophones / Agricultures*, N°10 (1), pp. 25-39.

INRA (1996), AGREV Rapport de synthèse, Département Systèmes Agraires et Développement, Unité Versailles-Dijon-Mirecourt, p. 41.

INRA (1997), Vittel, *Les Dossiers de l'Environnement de l'INRA*, 14, p. 78.

Istituto Nazionale di Economia Agraria (2000), "Le politiche comunitarie per lo sviluppo rurale". *Verso la nuova programmazione 2000-2006.*

Istituto Nazionale di Economia Agraria (2001), "Lo sviluppo rurale. Turismo rurale, agriturismo, prodotti agroalimentari", *Quaderno Informativo* n° 4.

Istituto Nazionale di Economia Agraria (2002), "La costruzione di percorsi di qualità per l'agriturismo*"*, *Quaderno Informativo* n°12.

University of Massachusetts Extension (2000), "What is Community Agriculture and How Does It Work?", www.umass.edu/umext/csa/about.html accessed August 2004.

Manktelow, D., T. Renton and S. Gurnsey, (2002), *Technical Developments in Sustainable Winegrowing in New Zealand*, paper given at the 2002 Romeo Bragato Conference.

Marsh, L.L. *et al.* (1996), "Introduction and overview" in L.L. Marsh *et al.* (eds.), *Mitigation Banking: Theory and Practice*, Island Press, Washington, D.C.

Mass, J. and J. Wisserhof (2000), "A pink invasion into the Dutch periphery", in R. Majoral, H. Jussila and F. Delgado-Cravidao (eds.), *Environment and Marginality in Geographical space: Issues of Land use, Territorial Marginalisation and Development in the new Millenium*, Ashgate Publishing Ltd., United Kingdom.

Mayo, T.D. (2000), "A Holistic Examination of the Law of Conservation Easements" in J.A. Gustanski and R.H. Squires (eds.), *Protecting the Land: Conservation Easements Past, Present, and Future*, Island Press, Washington D.C.

McKee, S. (2000), "Conservation Easement to Protect Historic Viewsheds: A Case Study of the Olana Viewshed in New York's Hudson River Valley", in J.A. Gustanski and R.H. Squires (eds.), *Protecting the Land: Conservation Easements Past, Present, and Future*, Island Press, Washington D.C.

Memon, A. and Perkins H.C. (eds.), (2003), *Environmental Planning in New Zealand*, Palmerston North, Dunmore Press.

Mie Prefecture Government website, www.pref.mie.jp accessed March 2004.

Nakata, T (2003), "A Study on the Volume and Transportation Distance as to Food Imports ("Food Mileage"), and its Influence on the Environment (in Japanese)", *Journal of Agricultural Policy Research*, No. 5, Tokyo, Japan

The National Trust, www.nationaltrust.org.uk, accessed March 2004.

The National Trust (2000), *Agriculture – 2000 and Beyond: An Agricultural Policy for the National Trust*, the National Trust, London.

Nelson, R. *et al.* (2004), "Natural Resource Management on Australian Farms", *ABARE e-Report* 04.7, Canberra, Australia.

Nestlé (2003), *Nestlé and Water, Sustainability, Protection, Stewardship*, Nestlé documents.

Netherlands,[Ministry of Agriculture, Nature and Food Quality, the Netherlands (LNV) (2001), *Manure and the environment: The Dutch approach to reduce the mineral surplus and ammonia volatilisation*, 2nd ed., The Hague, www.minlnv.nl/international/policy/environ/.

New Zealand Parliamentary Commissioner for the Environment, (2001) *Managing change in paradise: Sustainable development in peri-urban area*, Wellington.

New Zealand Wine and Grape Industry's website (2005), http://www.nzwine.com, accessed February.

Nobukane, T (2001), "*Supporting Agriculture in Collaboration with Producers and Consumers (in Japanese)*", Graduation Thesis for Hokkaido University, Japan.

OECD (1998), *Cooperative Approaches to Sustainable Agriculture*, Paris.

OECD (2001a), *Multifunctionality: Towards an analytical framework*, Paris.

OECD (2001b), *OECD Workshop on Multifunctionality: Applying the analytical frame-work*, http://www1.oecd.org/agr/mf/

OECD (2003a), *Multifunctionality: the policy implications*, Paris.

OECD (2004), *Tradeable permits: policy evaluation, design and reform*, Paris.

Oenema, O. *et al.* (1998), "Leaching of nitrate from agriculture to groundwater: the effect of policies and measures in the Netherlands", in Van der Hoek, K. *et al. Nitrogen, the Confer-N-s: First International Nitrogen Conference 1998*, Elsevier.

Oliver (1993), *The Oliver Report on the Constitution*, The National Trust, London.

Ondersteijn, O. *et al.* (2002), "The Dutch Mineral Accounting System and the European Nitrate Directive: Implications for N and P management and farm performance", *Agriculture, Ecosystems and Environment*, Vol. 92, Nos. 2-3, pp. 283-296.

Pearce, D.W. (1992), *The MIT Dictionary of Modern Economics*, Fourth edition, The MIT Press, US.

Peart, R. (2004), "*A Place to Stand: the protection of New Zealand's natural and cultural landscapes*", Auckland; Environmental Defense Society.

Perrot-Maître, D. and P. Davis (2001), Case Studies of Markets and Innovative Financial Mechanisms for Water Services from Forests, Forest Trends, Washington.

Primdahl, J. and S.R. Swaffield (2004), "Segregation and multi-functionality in New Zealand landscapes," in F. Blouwer (ed.), *Sustainable Agriculture and the Rural Environment,, Policy and Multifunctionality,* Edward Elgar, Cheltenham, UK.

Pruckner, G.J. (1995), "Agricultural Landscape Cultivation in Austria. An Application of the CVM", *European Review of Agricultural Economics*, N°22 (2), pp. 173-190.

Reibel, C. (1999), "Dans le périmètre de captage d'eau, 24 agriculteurs sous contrat avec Vittel", *Réussir Lait Elevage*, 117: 80-81.

Sasayama, T (2004), "*Community Supported Agriculture*", downloaded from his website, www.sasayama.or.jp/diary/CSA2.htm, accessed March 2004.

Shabman, L.A. *et al.* (1996), "Wetland Mitigation Banking Market", in L.L. Marsh *et al.* (eds.), *Mitigation Banking: Theory and Practice*, Island Press, Washington, D.C.

Shabman, L.A. (2004), "*Compensating for the Impacts of Wetlands Fill: The US Experience with Credit Sales*", in Tradable Permits: Policy Evaluation, Design and Reform, OECD, Paris.

Sivini, S. (2001). "Capo di S. Maria di Leuca Action Group – Italy", in Cavazzani A. and Moseley M. (editors), *The Practice of Rural Development Partnerships. 24 Case Studies in Six European Countries*, PRIDE Research Report, Rubbettino, Rende.

Small, S.J. (2000), "An Obscure Tax Code Provision Takes Private Land Protection into the Twenty-First Century", in J.A. Gustanski and R.H. Squires (eds.), *Protecting the Land: Conservation Easements Past, Present, and Future*, Island Press, Washington D.C.

Sohngen, B. (1998), "Incentive Based Conservation Policy and the Changing Role of Government", Working Paper for American Farmland Trust, www.farmlandinfo.org accessed May 2004

Studt, J. and R. Sokolove (1996), "Federal Wetland Mitigation Policies", in L.L. Marsh *et al.* (eds.), *Mitigation Banking: Theory and Practice*, Island Press, Washington, D.C.

Swaffield, S.R., and Fairweather, J.R., (2003) "Contemporary public attitudes to landscape", in *Reclaiming our Heritage,* proceedings of the New Zealand Landscape Conference 25-26 July, Takapuna Auckland, Environmental Defense Society, Auckland.

Thune, J.B. (2002) *The Flåm Railway*. Skald Publishers. ISBN 82-7959-029-3.

United Kingdom Government, *Explanatory note to Countryside and Rights of Way Act 2000*, www.hmso.gov.uk, accessed April 2004.

United Kingdom Government, Department of Environment, Food and Countryside Affairs website, www.defra.gov.uk, accessed March 2004.

United States Inland Revenue (2000), "Giving Charity by Individuals", Club and Charity Series IR65.

United States Natural Resources Conservation Service (2004), "National Resources Inventory: 2002 Annual NRI", www.nrcs.usda.gov accessed May 2004.

Verschuur, G., E. van Well and L. Boss (2003), *Study on the effect of selected EU environmental legislation on agriculture: The Netherlands*, Centre for Agriculture and Environment (CLM), CLM 565-2003, April, Utrecht, www.clm.nl.

Vliet, J. van and G. Ogink (2004), *Regeling Beëindiging Veehouderijtakken (RBV): Bijdrage aan de vermindering van het mestoverschot*, Expertisecentrum LNV, Ede (in Dutch).

Vukina, T and A. Wossink (2000), "Environmental policies and agricultural land values: Evidence from the Dutch nutrient quota system", *Land Economics*, Vol. 76, No. 3, pp. 413-429.

Waters, Tom (unknown), "Missouri Agricultural Mitigation Bank Pilot Project", www.mldda.org/wetlandbank.htm, accessed May 2004.

Weingartern, P. (2001), "Tradable pollution permits: A useful instrument for agri-environmental policy?", paper presented at the ACE Seminar on Environmental Effects of Transition and Needs for Change, Nitra, Slovakia, www.ceesa.de/NitraPapers/Weingarten.pdf.

White and Wild Milk website (2004), www.whiteandwild.co.uk, accessed April 2004.

Wiebe, K. *et al.* (1996), "Partial Interests in Land: Policy Tools for Resource Use and Conservation", *Agricultural Economic Report No.744*, Economic Research Service, U.S. Department of Agriculture, Washington D.C.

Wildlife Trusts website (2004), www.wildlifetrusts.org, accessed April 2004.

Wilson, H. (1992), *Banks Ecological region: Port Hills, Herbert and Akaroa Ecological Districts*: Protected Natural Areas programme Survey Report No 21, Department of Conservation.

Wossink, A. (2004), "The Dutch nutrient quota system: Past experience and lessons for the future", in OECD (2004).

Wossink, A. and G. Benson (1999), "Animal agriculture and the environment: Experiences from Northern Europe", paper presented to the Southern Extension Public Affairs Committee, Florida, http://www2.ncsu.edu:8010/unity/lockers/users/g/gawossin/Papers/Sepacf.pdf.

Wossink, A. and C. Gardebroek (2003), *The failure of marketable permit systems and uncertainty in environmental policy: The Dutch phosphate quota program*, http://www2.ncsu.edu:8010/unity/lockers/users/g/gawossin/quota.pdf.

Wossink, A. and F. Wefering (2003), "Hot spots in animal agriculture, emerging federal environmental policies and the potential for efficiency and innovation offsets", *International Journal of Agricultural Resources, Governance and Ecology*, Vol. 2, Nos. 3-4, pp. 228-242.

OECD PUBLICATIONS, 2, rue André-Pascal, 75775 PARIS CEDEX 16
PRINTED IN FRANCE
(51 2005 16 1 P) ISBN 92-64-01448-9 – No. 54511 2005